The Ethics of Cryonics

Francesca Minerva

The Ethics of Cryonics

Is it Immoral to be Immortal?

Francesca Minerva
Philosophy and Moral Sciences
University of Ghent
Ghent, Vlaams Brabant, Belgium

ISBN 978-3-319-78598-1   ISBN 978-3-319-78599-8 (eBook)
https://doi.org/10.1007/978-3-319-78599-8

Library of Congress Control Number: 2018939423

© The Editor(s) (if applicable) and The Author(s) 2018
This work is subject to copyright. All rights are solely and exclusively licensed by the Publisher, whether the whole or part of the material is concerned, specifically the rights of translation, reprinting, reuse of illustrations, recitation, broadcasting, reproduction on microfilms or in any other physical way, and transmission or information storage and retrieval, electronic adaptation, computer software, or by similar or dissimilar methodology now known or hereafter developed.
The use of general descriptive names, registered names, trademarks, service marks, etc. in this publication does not imply, even in the absence of a specific statement, that such names are exempt from the relevant protective laws and regulations and therefore free for general use.
The publisher, the authors, and the editors are safe to assume that the advice and information in this book are believed to be true and accurate at the date of publication. Neither the publisher nor the authors or the editors give a warranty, express or implied, with respect to the material contained herein or for any errors or omissions that may have been made. The publisher remains neutral with regard to jurisdictional claims in published maps and institutional affiliations.

Cover credit: Pattern adapted from an Indian cotton print produced in the 19th century

Printed on acid-free paper

This Palgrave Pivot imprint is published by the registered company Springer International Publishing AG part of Springer Nature.
The registered company address is: Gewerbestrasse 11, 6330 Cham, Switzerland

This book is dedicated to the memory of the people I miss and I wish I could see again: my aunts Masaria and Wanda, my grandparents Biagio, Liliana, and Saro, and family friends Cosimo and Cristina.

Preface

For the sake of transparency, I should clarify that, while I have not signed any personal arrangements to have myself cryopreserved, I have been considering the option for a number of years. In fact, I began writing this book as a means of consolidating and balancing my own views on a topic that has fascinated me for as long as I can recall. As a result, its argumentative structure strongly reflects my own indecision on the matters at stake, and my personal ambivalence towards indefinite life extension more generally. Most paragraphs start by presenting a certain argument against cryonics, and then continue by analysing counterarguments, counterarguments to those counterarguments, and so on. Unlike many cryo-enthusiasts, I have made a conscious effort not to dismiss objections to cryonics by overestimating the great potential of future technologies, and I am not excessively optimistic about future scenarios. I have also taken great care to avoid leaning too far in the opposite direction, as cryo-sceptics tend to do, by not brushing aside the possibility that cryonics might work one day, nor conflating what is unusual with what is immoral.

With this book, I have attempted to outline morally relevant aspects pertaining to cryonics, both as a medical procedure and as an intermediate step towards life extension. I have analysed moral issues about cryonics-related techniques and indefinite life extension, focusing on how they impact the individuals who undergo cryonics, the society they leave behind at the time of cryopreservation, and the society they are (potentially) revived into. At the same time, this book aims to be free from both the exceeding enthusiasm of some cryonicists and the disdain of some cryo-sceptics. Hence, what this book will *not* try to do is to convince anyone that cryonics is good or bad, nor to persuade anyone that they themselves should sign up for cryonics or cancel any cryonics arrangements they may already have. Moreover, this book will not offer a comprehensive introduction to cryonics in general, but only to the ethical aspects of cryonics and indefinite life extension. For this reason, non-ethical issues, such as technical aspects of the procedure, will be introduced only when relevant to the ethical analysis at hand.

While this book is obviously relevant to people interested in undergoing cryonics upon legal death, it should not be relevant only to them. Thinking about cryonics and life extension leads us to think philosophically about the life we are currently living, not just about the one we might get to live if cryonics works, and also about the philosophical meaning of life and death.

Contemplating the pros and cons of cryonics prompts a reflection on the circumstances under which we would accept to be cryopreserved, or about the circumstances under which we would want to be revived, in turn forcing us to think about what makes our life worth living right *now*. When confronted with the possibility of being revived after cryonics, only to find out that some of the information stored in our brain has been damaged or lost in the process, we cannot help but think about the very nature of that information—that unique combination of personality traits, memories, moral values, and desires that must somehow be maintained over time in order to rightfully say "*I* made it, *I* survived".

Whatever constitutes a reason for wanting to be revived after cryonics, or for giving one more shot at life through cryonics, is also what motivates us to keep going when faced with life's biggest challenges. Do we want to keep living out of curiosity, or out of hope about the future? Out of fear of death, or love of life? Out of a drive to finish what we've started, or because of the love we feel towards someone? By putting a relatively high monetary price on an extremely low chance of adding a potentially infinite number of years to our (future) life, cryonics sets a fascinatingly complex framework for a philosophical exploration of the meaning and value of life and death.

In the context of more immediately salient social matters, cryonics may even hold some promise as a sword to cut a Gordian knot. Presented as an alternative to abortion and euthanasia, cryonics offers a new perspective on such divisive issues by suggesting that, while profound disagreements between people holding irreconcilable moral values might not be solved, they might nevertheless be circumvented.

Ghent, Vlaams Brabant, Belgium   Francesca Minerva

Acknowledgements

This book would not exist, or would surely be a much worse one, if I had not received the feedback and support of some truly wonderful people.

The greatest debt of gratitude is towards Adrian Rorheim. He really did an incredible job proofreading and editing the manuscript. Not only did he turn my dry, non-native English academese into something much more pleasant to read, but he also provided me with interesting suggestions, comments, and feedback. He also proved to be incredibly patient dealing with my bouts of anxiety and last-minute requests, so I can't stress enough how profoundly and sincerely grateful I am for the job he did.

I would like to thank the welcoming, kind, and supportive community of cryonicists I found online. People say that the success of cryonics depends on the community of cryonicists; if that is the case, I am confident that cryonics will succeed. In particular, I received highly detailed and useful feedback on the first two chapters from Dr Mike Perry at Alcor, from Dr Aschwin de Wolf (also at Alcor), and from Mr Christopher Gillet. They have been incredibly generous with their time, and they have helped me achieve a result that is far better than I had initially hoped.

I would like to thank Prof. Tom Buller and Dr Ole Martin Moen for their very helpful and insightful comments on Chap. 3.

I would like to thank the Research Foundation Flanders (FWO) for supporting my research with a postdoctoral grant, and especially my supervisor Prof. Johan Braeckman for his constant and valuable support and help.

I am also grateful to my editors at Palgrave, Rachel Krause and Kyra Saniewski, for their patience and their precious help.

I would like to thank Dr Anders Sandberg. Over the years, we have talked at lenght about cryonics, and it is largely thanks to him that I felt confident enough to write this book (indeed, we co-authored two papers on this topic and are planning to write a third one). Few people have had a more considerable influence on this book than him, and for that I am forever grateful.

I am also grateful beyond words to my husband and "partner in crime" Alberto for his detailed, insightful, and immensely helpful feedback on the whole manuscript—not to mention for his support and encouragement throughout the writing of this book. As usual, he has been extraordinarily patient, understanding, and loving.

Finally, I want to thank my family for *not* caring that much about this book. That may sound sarcastic, but I mean it with the utmost sincerity. It often feels as though our value as individuals depends on our intellectual achievements. Being reminded that that is not the case—and that we are loved no matter what we achieve—is one of the highest privileges a person can enjoy.

Contents

Part I	**Cryonics as an Ethical Problem**	**1**
1	**Pausing Death**	**3**
	Weird Goals	3
	Starting and Ending Life in Liquid Nitrogen	7
	The Information-Theoretic Criterion of Death	9
	Public Scepticism Towards Biotechnology	12
	Against Nature	13
	Humans Should Not "Play God"	14
	Weirdness and Repugnance	15
	Uncertainty	16
	Only the Rich Will Be Able to Afford It	18
	References	21
2	**Resuming Life**	**23**
	Objections to Cryonics	24
	Waste of Resources	25
	Waste of Organs for Transplants	25
	Waste of Money That Could Be Used for Donation to an Effective Charity	27
	Indifference of the Future	34
	No Interest in Spending Resources on Reviving the Cryopreserved	35
	No Interest in Developing Cryonics Technology	36

    No Interest in "Homo sapiens"   37
    Desirability of Being Revived in the Future   38
        Trouble Adapting Even to an Objectively Better World   39
    References   42

Part II Cryonics as a Step Towards Immortality   45

3 The Death Conundrum   49
Is Death Bad?   49
Death as Transition to Nonexistence   50
    A Life Worth Starting and a Life Worth Living   51
    Whose Nonexistence?   53
Death as Deprivation   55
    The Harm of Deprivation   55
    The Plausible Counterfactuals   57
    Epistemic Disagreement About Plausible Counterfactuals   58
    Death as Deprivation of Negative Counterfactuals   60
Death as Frustration of Desires   61
References   66

4 The Immortality Conundrum   67
Different Types of Immortality   67
What Would an Indefinitely Long Life Look Like?   70
    Freedom from Regrets   70
Personal Identity   74
A Recognizably Human Life   78
Would Virtual Immortality Deprive People of Eternity
in Heaven?   83
Boredom   86
Tiredness   90
References   93

Part III Alternative Uses of Cryonics   95

5 Cryothanasia   97
   Objections to Euthanasia Applied to Cryothanasia   100
   Deontological   101
   Faith-Based   102
   Principles of Medical Ethics   103
   Weirdness and Repugnance   105
   Unlikelihood and Futility   106
   Resource Use   108
   References   109

6 Cryosuspension of Pregnancy   111
   Giving Pregnant Women Another Option   113
   *Would Objections to Abortion Apply to Cryosuspension
   of Pregnancy?*   115
     What Type of "Future Like Ours"?   116
     Potentiality   117
     Killing an Innocent Is Always Impermissible   119
   Reproductive Technology   120
   Therapeutic Aid   122
   Adoption   123
   Ectogenesis   124
   Limits   127
   References   130

Index   133

PART I

Cryonics as an Ethical Problem

INTRODUCTION

In this first part of the book, after briefly introducing how cryonics works and what kind of results it is expected to achieve, we will focus on the most common and most probable objections to cryonics. In Chap. 1, we will tackle the objections that are often raised against new biotechnologies in general when they are first introduced. Some of these objections deal with concerns about the fact that such new technologies are against nature or against God's will or just plainly weird and yucky. Usually, after some time has passed by and a new biotechnology is no longer perceived as "new", opposition based on these objections becomes weaker.

Some other objections to new technologies stem from epistemic uncertainty about the impact they might have, either on the individual who uses them or on society at large. For instance, some are concerned that cryonics, even if successful, could cause severe and permanent brain damage to cryopreserved individuals, so that they might be unable to actually benefit from being brought back to life. Other objections based on uncertainty focus on the negative impact that such technology could have on society; for instance, given the current cost of cryonics, some fear that it will worsen the socio-economic disparity between those who can afford it and those who cannot.

In the second chapter, we will focus on objections that pertain to cryonics more specifically. For instance, cryonics is often considered a waste of organs that could be donated and transplanted instead of being stored in cryonics facilities, and some also consider it a waste of monetary resources that could be donated to effective charities in order to save the lives of poor people.

Finally, we will consider objections based on the idea that cryonics has zero chance of success. One of these objections is based on the idea that future generations will not have an interest, or will not have the resources necessary to revive the cryosuspended. The other set of objections is based on the assumption that people who would eventually be revived would never adapt to their new lives, and they would regret having invested money on cryonics.

CHAPTER 1

Pausing Death

Abstract Cryonics is the act of preserving legally dead individuals at ultra-low temperatures, in the hope that they can someday be revived using future technology. Although still in its infancy, the potential success of cryonics carries many crucial implications for human society, and discussing these ahead of time may help us avoid unwelcome developments and unnecessary conflicts. However, the discussion around cryonics has not advanced significantly since its introduction over 50 years ago, and we must look to other advances in biotechnology for clues about how society should deal with cryonics. This chapter reviews the public response to in vitro fertilization (IVF) and embryo cryopreservation (EC), two closely related medical technologies that have been available to the public for over 30 years and that share many relevant aspects with cryonics. Objections to IVF and EC have tended to fall into five broad categories: unnaturalness, playing God, repugnancy, uncertainty, and social inequality.

Keywords Cryonics • Cryopreservation • IVF • Biotechnologies • Medical ethics • Bioethics

Weird Goals

Whenever cryonics appears in the media, it is nearly always portrayed with a certain dose of scepticism, and is paired with such adjectives as "weird", "crazy", "fictional", or "scammy". For instance, a 2015 article in the

Financial Times was titled "Inside the weird world of cryonics", a headline not too different from that of a 2016 *National Post* article entitled "Inside the strange world of cryonics, where people are 'frozen in time' with hopes of escaping death". As anyone can independently verify through a quick Google search, similar headlines abound in the hundreds. Typically, the tone of the article is one of incredulity, often with a hint of derision and sometimes even outright contempt. The message usually goes along the lines of "there are some weird people planning to do this strange thing in order to achieve a crazy goal."

To be fair, the plan behind cryonics—also known as *cryopreservation* and *cryosuspension*—is quite ambitious. In short, the idea is that one can preserve legally dead individuals in liquid nitrogen and store them for decades or even centuries, in the hope that future technology will not only succeed in reviving them with their mental faculties intact, but also cure the condition that led to their demise in the first place. Among people choosing to undergo cryonics (henceforth referred to as *cryonicists*), the most optimistic ones hope that they, once revived, will have the option of undergoing various rejuvenating treatments, allowing them (at least in theory) to live indefinitely, or at least for many more years than they would have otherwise lived. As of 2018, cryonics companies already exist in the United States and Russia. Although countries differ in their legal requirements for allowing a citizen to be cryopreserved, the option is generally open to any autonomous, consenting adult willing to pay the fee required by the cryonics provider.

So cryonics might not be the ideal icebreaker on a first date, so to speak. But weird goals are not necessarily bad or irrational goals, and investing in an unusual plan does not make anyone a bad person. Yet some of the moral objections to cryonics seem to suggest that, at the heart of the disdain towards cryonics, there may be a conflation between what is considered weird or unusual and what is considered immoral.

In this sense, scepticism towards cryonics should not come as a surprise. Most novelties in the biomedical sciences have elicited, and continue to elicit, similar reactions.

In general, it seems that the closer a technology is to interfering with issues of life and death, the stronger the scepticism or outright aversion it raises. As an illustration of this fact, it is both helpful and interesting to look at the historical change in attitude towards another kind of technology that was also considered "weird" and "immoral" at first, but that is now widely accepted—namely in vitro fertilization (IVF) with embryo cryopreservation (EC). Despite having faced considerable initial scepticism after their introduction in 1983, IVF and EC have since found their place in the public's sphere of acceptance and become part of standard medical practice in most countries.

Cryonics has not become as popular and is not perceived as "normal" as IVF and EC, even though it has been discussed for a very long time.[1] In fact, the first real case of human cryopreservation took place in 1967, thus predating the birth of Louise Brown, the first IVF baby, by 11 years. Although the procedure was primitive by today's standards, the patient—psychology professor James H. Bedford—remains in cryosuspension to this day.

There are several reasons why IVF and EC have fared much better than cryonics. The main reason is probably the technical complexity of cryonics: it requires far more advanced technology and knowledge to cryopreserve a fully developed person, made up of trillions of highly specialized cells, than to preserve an embryo consisting of just a few cells. But even if the cryopreservation process itself were simpler, the revival process remains incredibly difficult to realize. More than half a century after the first cryopreservation, no attempt to revive a cryopreserved patient has been made, and there is no agreement among experts about whether and when revival will be feasible. IVF and EC, on the other hand, produced reliable results soon after they were first developed, allowing them to be made publicly available rather quickly. As a result, their increasingly common use has made them look less weird and uncontroversial. Given the lack of confidence in the possibility that revival will someday become available, it is easy to understand why people choose not to invest in cryonics and keep perceiving it as weird.

But technical difficulties alone do not explain the unusual deadlock of cryonics compared to other technologies, as cryonics is surely not the first extremely ambitious plan in the history of humanity. A century ago, it would have seemed ludicrous to suggest that humans might someday visit another planet in the solar system. Since then, humans managed to land on the Moon, and now the prospect of visiting and even colonizing Mars is on the table.

One reason why there has not been as much progress in the research on cryonics is a lack of funding. Cryonics research is not supported by state funding, but is instead conducted largely by private research groups relying on private donations. This lack of funding hampers progress, and this lack of progress is, in turn, used as a reason for not investing in more research. Needless to say, one cannot be sure that more research would guarantee success; but at least it would allow a more accurate assessment of the potential of cryonics, and make it a bit less dubious whether it is a potentially successful enterprise.

In the first chapter of this book, cryonics will be frequently compared with IVF and EC. Despite the obvious anatomical differences and the (perhaps less obvious) moral differences between embryos and fully developed humans, there are a number of morally relevant similarities in how cryopreservation is applied to each of the two. For example, both practices essentially aim at interfering with natural processes—embryonic development in the case of EC and death in the case of cryonics—by using ultra-low temperatures to slow down the body's metabolic activity.

EC and IVF carry the unique advantage of having been introduced recently enough for many readers to have witnessed first-hand the change in society's attitudes towards the two. It took only three decades for EC and IVF to go from being seen as weird and immoral to being considered normal and, by many, good. People today who are against cryonics because it is weird need only look at the recent acceptance of EC and IVF to see that weirdness alone is a bad proxy for moral permissibility. What this consideration suggests is that, *if* cryonics is unethical, it must be unethical for reasons that have nothing to do with its being weird and unusual. Instead, we would need to ask whether it causes harm to individuals and/or societies, either now or sometime in the future.

Comparing cryonics to EC is also useful because it shows the enormous potential of cryopreservation beyond IVF and cryonics. In Chap. 6, we will discuss one instance of such potentiality, namely a hypothetical future technology aimed at cryopreserving human foetuses. At the moment, it is not possible to cryopreserve embryos beyond the blastocyst stage, which occurs roughly five days after fertilization, that is, when the embryo is five days old. Meanwhile, we are as far from being able to preserve foetuses as we are from being able to preserve adult humans—indeed, we are not even able to keep a foetus younger than 24 weeks alive outside the womb. We will see how cryonics, paired with hypothetical techniques that would make it possible to extract the embryo/foetus without causing damage to its tissues, could become a less controversial alternative to abortion. Given that abortion is regarded as one of the most divisive social issues in many cultures, this option would not only help women who cannot continue their pregnancy and yet do not want to abort, but would probably also help reduce the conflict between pro-life and pro-choice groups by offering a best-of-both-worlds compromise solution.

On a similar note, we will also explore the potential of cryonics as a practical alternative to euthanasia, another highly controversial medical procedure. Although euthanasia is illegal in most countries, there is a growing global movement in support of its legalization. As we will discuss in Chap. 5, so-called cryothanasia may offer a less permanent alternative to euthanasia for patients suffering from prolonged, unbearable, and incurable pain. Unlike euthanasia, cryonics does not seek to end someone's life, but rather to pause it in the hope that future medical technology will give them a new chance to continue living.

So perhaps some technologies, including cryonics, can be used to finally overcome profound disagreements in our societies. And although such technologies often provide grounds for new conflicts between, say, the religious and the atheist, or the conservative and the liberal, we will see how technology may also be used to mend social fractures that originate in non-negotiable and irreconcilable moral views.

Starting and Ending Life in Liquid Nitrogen

Over the past two centuries, science and technology have progressed at an extraordinary speed. For better or worse, humans have gained the knowledge necessary to understand and, to a certain extent, to alter some of the most complex mechanisms governing the natural world.

One of the many groundbreaking achievements in recent decades has been the development of reproductive technology, allowing ever more control over the earliest stages of the human life cycle. Even though this kind of technology receives less public attention than many other recent feats, it has had a significant impact on society: since 1978, around 5 million people have been born thanks to in vitro fertilization (IVF). In 1983, it became possible to also cryopreserve embryos in liquid nitrogen at a temperature of −196 °C, thereby enabling prospective parents to implant their embryos long after they were conceived in the laboratory (Andersen et al., 2005; Horsey, 2006).

Given how IVF and embryo cryopreservation (EC) tend to be last-resort options for parents struggling to conceive, it is likely that most of these children would not have been born if these options had not been available. IVF was also the first successful attempt at going beyond the traditional medical aim of "merely" saving lives by actually *conceiving* life through unnatural means, which until then had largely been considered a prerogative of gods or nature.

Modern medicine and technology have achieved extraordinary results in advancing the knowledge of the human body, restoring health, and increasing both the quality and the length of our lives. But what if medicine and technology succeeded not just at conceiving life, healing diseases, and stretching the human lifespan, but also at increasing our capacity to control the other end of the spectrum of life? What if we could bring back the dead (in the specific sense of "dead" I will specify below) to the world of the living?

Cryonics—also known as *cryopreservation* or *cryosuspension*—is the act of preserving legally dead individuals at ultra-low temperatures, typically using liquid nitrogen. Such extremely low temperatures can, in effect, "pause" metabolic processes to a point where the body is completely inactive and does not decompose, making it possible—at least in theory—to "un-pause" them at a later time. The hope is that, in the future, it will be possible to revive cryopreserved individuals and recover their body, memories, and personality (Minerva & Sandberg, 2015).[2]

As one cryonics research group explains on its website, a person is only really dead "once the chemistry of life becomes so disorganized that normal operation performed by a human body cannot be restored" (Alcor, n.d.).

Over time, medicine and technology have elevated the bar for what constitutes unfixable disorganization in the chemistry of life. Mere decades ago, cardiac arrest was considered a lethal event, believed to cause near-instantaneous loss of all information in the brain. Nowadays, a person is only considered dead around six minutes after the heart has stopped beating, as we now know it is only after the six-minute mark that brain death actually occurs. But the goalpost keeps inching forward, and it is not unreasonable to suppose that future technologies will allow us to buy even more time between the moment the heart stops beating and the point of irreversible brain death. As technology advances, death retreats.

The circumstances under which an individual is doomed to die keep shifting at the pace of technological advance. A good illustration is the increasingly early stage of pregnancy at which a foetus is considered viable. Only 30 years ago, a 24-week-old foetus would not have been considered viable. It would have been doomed to die within hours of being born, with no attempts made to resuscitate it and keep it alive. Nowadays, even though most of the preterm babies born at 24 weeks frequently experience lasting developmental issues compared to their full-term peers, they nevertheless tend to survive (Younge et al., 2017). The history of medicine is filled with similar examples, and it seems we are only as doomed to die as the technology we need to survive is lacking. At the moment, technology is not sufficiently advanced to allow us to survive as long as we please. Hence, the best we can do is use cryonics to try to halt the process of dying, in the hope that future medicine will fix what is killing us today.

The Information-Theoretic Criterion of Death

The criterion for declaring someone dead has been modified over the years, and it is not the same across different countries or religious views. Differences in the chosen criterion of death often have to do with the kind of technology available or with policies regulating the use of organs for donations (Sade, 2011).

For instance, the *cardiopulmonary standard* of death relied on the lack of cardiac and respiratory activities to certify the death of a patient. However, with the development of machines able to perform these functions artificially, such as ventilators, this definition quickly became obsolete. The *whole-brain standard* of death, which is the most widely accepted in current medical practice, defines death as the irreversible cessation of all brain functions, as individuals whose whole brains have stopped functioning have lost the capacity for consciousness as well as autonomous respiration. In recent years, the whole-brain standard of death has also been subject to criticism, partly because of new findings on patients with *locked-in syndrome*. These patients are conscious but completely paralysed, with the only exception of the eyes (in some cases). Some of them, however, do not show significantly more brain activity than brain-dead individuals—and yet they are obviously not dead, as they are conscious. It has therefore been suggested that a more adequate standard of death should refer to the loss of the capacity for consciousness, memory, beliefs, desires—in sum, to the loss of capacity of performing activities we consider specifically human (McMahan, 1995). According to this new criterion, called the *higher-brain standard*, death occurs when one becomes irreversibly unconscious.

Cryonicists share a different definition of death, referring to the *information-theoretic criterion*. The concept of "information" is used here in a very broad sense, such that also consciousness and self-consciousness can be considered as particular types of "information". According to the information-theoretic criterion, death occurs only when the information stored in the brain becomes so corrupt that it would be impossible to retrieve it. At that point, there is no longer any chance of recovering its unique information—including consciousness and self-awareness—by any current or future technology. According to this definition of death, as long as information within an individual's brain is not corrupted to the point that it could never be recovered, one is not irreversibly dead—because, at least in theory, it could be possible to retrieve the information stored in their brain.

Before we delve deeper, let us first consider just what it means to be alive or dead in the information-theoretic sense. For this, we need a bit of background on what it means to be alive, to have a mental life and an identity, and to be conscious.

The dominant view on human consciousness among cryonicists holds that what we commonly refer to as a "person" is essentially a unique collection of information stored inside a brain.[3] This information includes everything from basic instincts and congenital quirks, through subconscious biases and preferences, and all the way up to cherished memories, defining experiences, learned skills, moral and political views, novel ideas, and so on. These are all the qualities that, when put together, define the unique identity of one human being. The brain itself, meanwhile, is an extremely complex organ that encodes, stores, and employs all of this information in a concerted effort to ensure the survival and reproduction of the genes from which it grew. Although we have quite a lot of information about the detailed workings of individual systems within the brain—chemical pathways, cells, tissues, and so on—we know very little about how it all conspires to store vast amounts of detailed information about the world, and practically nothing about the conscious experience that comes with it. According to cryonicists, what we do know is this: one aspect that makes a given person that particular person is the information stored in (and processed by) their brains (De Wolf, 2015). We might dispute whether this is a sufficient condition for personal identity per se, but it seems uncontroversial that it is at least a necessary condition. For example, we often say that people with severe dementia who no longer possess relevant information about their past "are no longer themselves".

In order to better understand the difference between death with and without cryopreservation, let us for a moment imagine ourselves as very advanced laptops. After all, laptops also store and process information, and that information is at least unique enough to concern us if our own laptop were suddenly replaced with someone else's. Even though these similarities are surely not strong enough to claim that we are just like computers, they are not trivial enough for this comparison to be discarded as absurd either.[4]

Now, suppose that someone were to throw my laptop into the lava lake of an active volcano, much like the fate that befell Sauron's Ring of Power at the hands of Frodo Baggins in Tolkien's famous book. After mere seconds in the scalding lava, the content stored in my laptop (assuming I have been so foolish as to have made no backup) is irreversibly lost, its storage unit destroyed by the intense temperature. This is, in effect, what happens after a person dies and their bodies are cremated or buried: the information stored in their brain is destroyed, whether through incineration or decomposition, and hence irretrievably lost.

Imagine now a second scenario, in which a malfunction causes the lithium-ion battery in my laptop to spontaneously catch fire. Although I grab a fire extinguisher and manage to put out the fire within seconds, the sudden, intense heat severely damages the laptop's internal components, including its precious storage unit. Upon bringing it to a repair technician, I am informed that while the storage unit is badly burned, the information encoded on it may be largely unharmed. Unfortunately, the technician lacks both the tools and the skills needed to extract the information without destroying it in the process, and he does not know of any other technician who might be up to the task. However, he reassures me that there is great demand worldwide for a workable solution to this kind of problem, and that future engineers will probably find a way to retrieve the information in my laptop storage.

In the case where my laptop is thrown into the volcano, I would cry: "My laptop is dead!" and mourn the irretrievable loss of my precious data. But in the case where the battery of my laptop caught fire but the storage unit did not get completely destroyed, experts might one day be able to retrieve my data and transfer it to a new laptop. If computers will ever be self-conscious, the "information" experts might be able to retrieve might include such self-consciousness (remember that I am using the concept of "information" very broadly). While this laptop would be technically distinct from the one that caught fire, it would, for all intents and purposes, be equivalent in terms of information contained to the one I had, because all (or at least a large part of my files) would be retrieved and uploaded. The hope of cryonicists is either that the information stored in their brain could one day be retrieved and transferred to another substrate so as to preserve their identity (e.g. through so-called brain-uploading, which, however, will not be discussed in this book) or that the original "laptop", that is, their own original body and brain, could one day be revived with all its original information.

At the moment, only a few hundred people are cryopreserved, and no attempt has been made to revive any of them so far. This is because the technology required to revive them is not yet available, and we do not know if it will ever be. However, as is often the case, the theoretical discussion of a technological innovation precedes its actualization, and the debate about cryonics and its implications has been going on for decades despite the technology itself being far from fully developed or available. At this point, if concerns raised are outstandingly negative, the development of the technology could be paused or even stopped. This is what happened, for instance, with human cloning technology: when the birth of Dolly the sheep, the first mammal to be cloned, was announced to the world on 3 February 1997, the public soon became preoccupied with the possibility that such technology would someday lead to attempts at cloning humans. Even though there was no short-term plan to clone a human being, the concerns expressed by a large part of the public and by many experts led to a ban on human cloning in many countries around the world (Pattinson & Caulfield, 2004).

Public Scepticism Towards Biotechnology

It is not surprising, then, that the debate around the ethical implications of cryonics dates back as early as 1962, when Robert Ettinger—later known as "the father of cryonics"—first presented the idea in his book *The Prospect of Immortality* (Ettinger, 1962). Since then, the debate has proceeded rather slowly, owing perhaps to its widespread perception as unfeasible and remote—in contrast to IVF and human cloning, both of which have tended to produce more tangible results—but possibly also due to cryonics researchers being stigmatized for their work (Topping, 2016). Just as EC was initially met with suspicion and fear, the prospect of human cryosuspension after legal death is not usually viewed with much warmth and favour. Indeed, if anything, it is generally met with even greater hostility and/or incredulity than IVF was.

These days, IVF and EC are largely accepted. But when they were first introduced, most people were either sceptical of or outright opposed to the use of such technologies as a means to bring new humans into the world (Singer & Wells, 1983). We will now consider some of the most common arguments used against IVF and many new biotechnologies in general, as they provide a good indication of what arguments might be used against cryonics qua technology that attempts to interfere with something as central to humans as death.

Against Nature

One popular objection to IVF, also used against many other new technologies when they are first introduced, is based on the argument that we have a moral duty to follow nature. This argument, in turn, stems from the belief that we can, by observing how things *are*, deduce how things *should be* (a textbook case of what is known as the *naturalistic fallacy*). According to this argument, since humans have always reproduced through sexual intercourse, they also ought to continue to reproduce only through sexual intercourse, because there is something intrinsically good in natural processes.

Before discussing what is wrong with this type of argument, it is interesting to note that the force of arguments based on "nature" often seems to depend on the perceived "novelty" of a certain technology, that is, on whether and to what extent people are accustomed to it, rather than on whether the technology actually goes against what is perceived to be "natural". For example, this argument seems to hold much less force now, 40 years after the first IVF baby was born, than it did when IVF was first introduced to the public. These days, the technology has largely lost its novelty in the public eye, and so naturalistic arguments are often seen as outdated. In general, such arguments are often abandoned after some time, as people gradually become more accustomed to the technology in question. Eventually, something newer appears on the horizon, and the original technology ceases to be considered novel and controversial.

We can see how the same argument could be used against cryonics: to die is part of the natural cycle of life, hence we ought to die; therefore, trying to cheat nature by using preservation in liquid nitrogen to pause the process of dying must be immoral.

There is, however, at least one major problem with arguments based on the supposed moral superiority of the natural over the unnatural: the difference between natural and unnatural is far less obvious than one might think. As J.S. Mill (1874/2009) explained: "Nature in the abstract is the totality of the powers and properties of all things. 'Nature' means the sum of all phenomena, together with the causes that produce them; including not only everything that happens but everything that could happen."

If, as Mill suggests, "nature" is the sum of all phenomena and their causes, then everything—including humans and their technology—must be part of nature. What is produced by humans, and is commonly defined artificial, cannot be produced by anything but natural forces. In other words, what we define as artificial is not the creation of something new, but merely the rearrangement of what is already found in nature, according to nature's laws: "[T]he role of man is a very limited one; all we do is to move things into certain places. By moving objects we bring separated things into contact, or pull adjacent things apart; such simple changes of place produce the desired effect by bringing into play natural forces that were previously dormant" (ibid.). If everything is necessarily natural, then IVF and cryonics must be natural, too.

One could reply that "natural" refers to anything that was not created by humans. This presupposes that humans are the only beings in nature with the power to transform what is natural into what is artificial. According to this view, then, things like oceans, supernovae, and the Ebola virus are natural, whereas things such as lasagna, houses, and antibiotics are unnatural.

The problem with this distinction between natural and artificial is that, when used as an argument to support the superiority of the former over the latter, it necessarily winds down to extreme or inconsistent views. Since even the simplest forms of human technology would be "unnatural" under this definition, living "naturally" would mean eschewing such basic things as clothing, buildings, agriculture, and even simple tool use. One would have to live naked in a cave, eating only wild fruits and seeds and whatever small animals they managed to catch with their hands. Given such extreme implications, it is understandable why supporting a similar view leads almost necessarily to inconsistencies. The same people who would oppose IVF or cryonics on the basis that they are not natural tend, like most of us, to enjoy living in the comfort of their houses and rely heavily on the use of their personal technology.

Moreover, to say "all that is natural is also good" is plainly false. Tsunamis are entirely natural phenomena, yet we would not say that the 2004 Indian Ocean tsunami, which killed up to 280,000 people, could be considered good.

Humans Should Not "Play God"

Another common form of objection to new reproductive technologies, and biotechnology in general, is based on the argument that "we should not play God." This objection is often applied to human interventions that appear to deviate from the plan or will of some religious deity. Like objections based on the naturalistic fallacy, the "we should not play God" objection tends to get abandoned after enough time has passed and the fear of novelty has faded away. This kind of objection has been used throughout history in opposition to all sorts of things: eyeglasses, contraception, abortion, IVF techniques, euthanasia, cloning, and genetic engineering, to name but a few. Needless to say, tampering with death—as cryonics aims to do—also falls among activities considered to be God's prerogatives.

Much like the naturalistic argument, the "playing God" objection is used quite inconsistently by its advocates. Consistency dictates that both objections should also be used against many forms of life-saving medical treatments, spanning from antibiotics to intensive care units (ICUs); yet proponents of both naturalistic and "playing God" objections are generally silent on such matters.

Weirdness and Repugnance

New forms of technology are often seen as "weird", sometimes even "gross", presenting another class of arguments against IVF and other biotechnology. Such arguments are rarely given on their own but, instead, tend to be paired with arguments of the "unnatural" or "playing God" kind. According to this view, the fact that something "feels" weird or evokes a disgust reaction should be considered symptomatic of it also being immoral. For instance, philosopher Leon Kass (2001) has argued that:

> [R]epugnance is the emotional expression of deep wisdom, beyond reason's power fully to articulate it ... repugnance may be the only voice left that speaks up to defend the central core of our humanity. Shallow are the souls that have forgotten how to shudder.

So, consistently with this view, the fact that some people have a yuck reaction to the idea of suspending human bodies and disembodied heads in vats of liquid nitrogen for future revival should be considered proof that cryonics is immoral. But arguments based on gut reactions are hardly a good guide to morality.

Repugnance can be plainly irrational (I find olive pits repugnant, but this does not make olive pits immoral) and also heavily influenced by cultural influences or individual predispositions. For instance, the idea of transplanting a human heart—commonly portrayed as the seat of emotions—seemed both weird and repugnant when it was performed for the first time in 1967. Similarly, IVF was seen as weird and yucky for at least a decade after its introduction. And yet, few people today consider heart transplants or IVF "yucky". If we had given in to the gut reactions of the time, hundreds of thousands of people would have died because of diseases curable with organ transplants, whilst many others would not have been born at all because of infertility.

So if history is any indication, it seems that trusting our gut reactions is not wise. Indeed, if anything, we have learnt that disgust can be a very misleading guide to morality.

Uncertainty

Objections to new technologies based on uncertainty are very common. We easily get worried about the possible negative effects of a new technology.

Perhaps the best example is that of genetically modified organisms (GMOs). Soon after the first GMO tomatoes arrived on the market in 1994, the new products were met with numerous objections from the public. People were worried about GMOs causing all sorts of diseases, or claimed it was some sort of disgusting "Frankenfood" that would contaminate and then destroy all the natural, nutritious, good old-fashioned crops in the world. More than two decades and hundreds of studies later, it is clear that most GMOs, and certainly those present on the market, are not harmful to human health, have no noticeable difference in taste from other foods, and do not destroy natural crops (Nicolia, Manzo, Veronesi, & Rosellini, 2014).

Admittedly, though, arguments from uncertainty should not be discarded too lightly, as it is often difficult to predict the effects that a new medical procedure will have on the individuals undergoing it. Caution is always recommended, and adopting the precautionary principle might sometimes be the most rational strategy. Unfortunately, empirical uncertainty can only be overcome through experimentation. For instance, we know now that children born through IVF, with or without EC, are as healthy as children conceived naturally. Of course, given that the first person conceived through IVF, Louise Brown, is only 40, there is still a chance that IVF will turn out to have devastating epigenetic effects that only appear past the age of 40. However, this possibility currently seems vanishingly small, and it appears far more likely that people born through IVF are, indeed, just as healthy as naturally conceived ones.

Similarly, there is empirical uncertainty about the effects of cryopreservation of adults. We do not yet know whether this process would have any adverse effects or the extent and the nature of such effects. For instance, it is possible that cryopreservation would cause major disabilities, or permanent amnesia. The range of possible side effects is enormous, but it is also possible that cryopreserved people will, just like people born from cryopreserved embryos, turn out to be healthy. Whether or not we should proceed with cryopreservation in spite of this uncertainty depends on whether it is prudent, rational, and ethical to adopt a precautionary principle. Although we will not go into detail about the precautionary principle, it should be noted that if the principle had been rigorously applied to every new technology in the past, we would not have many of those technologies today (including IVF and EC). Perhaps this consideration is sufficient to justify proceeding with research on cryopreservation despite the risks, albeit with a cautious attitude.

Other objections to new reproductive technologies were based on uncertainty around the religious issue of when one's soul enters one's body, known as *ensoulment*. According to some religious traditions, notably that of the Roman Catholic Church, ensoulment happens at the moment of conception.[5] Given that fertilized eggs and embryos can be cryopreserved for a long time before being implanted and actualizing their potential to become a baby, a dilemma has arisen regarding what happens to the soul during the intervening time between the conception and implantation of the embryo. This question is particularly worrisome when one considers embryos that are conceived and cryopreserved, but never get implanted. Destroying these embryos might be considered morally equivalent to performing an abortion or a homicide, but leaving them in liquid nitrogen for decades or possibly centuries poses difficult questions about the status of their souls.

Cryopreservation of adults poses a similar problem: according to some religious views, when people die, their soul leaves the body and joins God in Heaven or some other spiritual dimension (if they have behaved well, that is). It is unclear what would happen to the soul if the individual were cryopreserved in view of someday being revived, instead of being buried or cremated. If the soul left the body before cryopreservation, then unless it somehow returns to the body from Heaven once the body is revived, it appears the revived person would find themselves without a soul upon revival.[6] If the soul instead remained attached to the body, but it turned out that revival of cryopreserved people is never feasible after all, the cryopreserved individual would be deprived of the chance to spend eternity in Heaven. Immortality, whether preceded by cryonics or not, in general poses some similarly difficult issues (we will return to this in Chap. 4); for example, if the soul is immortal and life on earth is only a brief transition towards spiritual eternity, then physical immortality opens a series of difficult issues.

Only the Rich Will Be Able to Afford It

A common concern about the development of new technologies is that since they are usually quite expensive upon commercialization, only rich people will be able to afford them, thereby causing the inequality between the rich and the poor to increase. This was indeed a concern also when IVF was introduced, because it was feared that only rich infertile people would be able to have children, whereas poor infertile people would not have the same option. Since having children is often a profound desire, and some people experience deep emotional distress because of their infertility, it is quite clear why the prospect of IVF raised some serious concerns with respect to fairness. IVF is now often supported by the public health system, and although it is usually not fully covered, it is generally affordable to anyone with the financial means to support a child in the first place.

But if inequality issues arise when it comes to procreation, they are more powerful when it comes to matters of death. If only the rich can afford cryonics, then being poor will be incomparably more disadvantageous than being rich, as it would suddenly make all the difference between living a normal lifespan and living, potentially, indefinitely.

One common counterargument is based on the fact that the price of new technologies tends to decrease over time. Computers, for instance, were originally very expensive, but nowadays one can buy a cheap one for a few hundred dollars. It is likely that cryonics, too, would become cheaper over time, following the usual declining cost trajectory of other (bio-) technologies.

Today, the cost of cryonics is not negligible, and it is surely not affordable by everyone. However, there are insurance companies that cover cryonics costs for fees between $30 and $100 per month. One cryonics provider, Alcor, offers full-body cryopreservation at a price of $200,000, while their competitor the Cryonics Institute charges around $28,000 for the same deal. Unlike Alcor, however, the Cryonics Institute does not include the cost of transportation to the cryonics facility, which can amount to as much as $90,000 (depending on the distance between the place where the person is declared legally dead and the cryonics facility). So all in all, if one decides not to pay the subscription upfront in favour of a monthly fee to an insurance provider, cryonics can already be afforded by not only the rich, but also people with a middle-class income.

Of course, people who live around or below the poverty line would not be able to invest $30–$100 in cryonics every month, so there would still be large parts of the population who could not afford it. The fact that so many people in the world still live in a state of poverty is, without a doubt, one of the greatest tragedies of our time; but while it is important to remain vigilant with respect to new sources of inequality and try our best to avoid worsening the current situation, focusing on just one potential source of inequality is not particularly useful. It is true that if cryonics were so expensive to be affordable only to very rich people, so that they would live much longer than poor ones, cryonics would end up increasing the current inequalities between the rich and the poor. However, there are many other factors that increase such inequalities. Compared with the poor, rich people have access to better healthcare and nutrition, and seem, on average, to be healthier overall. Indeed, rich people already live longer than poor people, and cryonics would only exacerbate this inequality without compensating for it in some other way.

Of course, to say that inequalities already exist does not imply that new sources of inequality should be taken lightly. On the contrary, since we know all too well that economic disparity is linked to differences in lifespan and overall well-being, we understand that they need to be taken very seriously. However, the point is that inequality is not per se a sufficient argument against the introduction of a new technology: morally, the problem is not that the technology will increase current inequalities, but that inequalities exist in the first place. The solution is not to eliminate technologies, but to address those socioeconomic circumstances by virtue of which the new technology would exacerbate inequalities. Besides, both in the case of IVF and in the case of cryonics, it seems that economic inequalities do not affect the access to these technologies in particular ways. As we said, IVF is largely subsidized with public money, making it accessible to middle and lower classes at the expense of taxpayers. For what cryonics is concerned, meanwhile, insurances provide a means to make it accessible to the middle class, albeit while still excluding the very poor. Over time, if more people sign up for cryonics, it is possible that costs will go down—eventually making it cheap enough for large parts of the population to afford it.

In this first chapter, we have analysed some possible objections to cryonics, starting from the most common objections to new technologies in general. We have seen how new technologies are often opposed because they are perceived as unnatural or against God's will or because they appear to be unusual or even "yucky". Arguments based on this kind of objections are often best explained by a status quo bias. Over time, as these technologies are utilized by more and more people, they stop being considered as a novelty and as unnatural, weird, against nature, and, ultimately, immoral.

We have also discussed concerns for the consequences that a new technology could have on the individual and on society at large. For instance, the uncertainty about the possible side effects of cryonics is sometimes considered a strong enough reason to reject cryonics on a prudential basis. As we have seen, this objection has some force, but it is doubtful that it is strong enough to undermine the cryonics project. Another concern is based on the fact that, when new technologies are introduced, they are usually quite expensive, hence accessible only to the rich. Poor people who would not have access to cryonics would be doomed to stay dead forever, experiencing yet another disadvantage; we have seen how this objection, too, is misplaced.

In the next chapter, we will focus on ethical issues that pertain to cryonics specifically. Thereafter, in Part II, we will tackle objections to cryonics understood as a step towards indefinite life extension.

Notes

1. Mike Perry reports that John Hunter, a renowned physiologist and surgeon, in 1776 attempted to freeze a fish with the idea that the process might be reversible.
2. For a discussion of a wider range of issues within cryonics, I highly recommend De Wolf and Bridge's edited volume of selected articles from *Cryonics Magazine*, entitled *Preserving Minds, Saving Lives* (Alcor Life Extension Foundation, 2015) as well as the official websites of Alcor (http://www.alcor.org/) and the Cryonics Institute (http://www.cryonics.org/).
3. The topic of consciousness is very complex and cannot be adequately discussed here. For a comprehensive overview, see, for example, Chalmers (1997).
4. Indeed, even though artificial intelligence currently lacks consciousness (widely considered a defining quality of humans), experts are discussing the possibility that machines may one day have the capacity for consciousness, and might therefore come to resemble humans in important respects. For more information, see, for example, Koch and Tononi (2008).
5. This is not an issue for all religions. For instance, according to Islam, the soul enters the body only 120 days after conception. Moreover, the Catholic Church originally shared the Aristotelian view that ensoulment happens at 40 days for male embryos and 80 days for females.
6. For an analysis of this issue, see, for example, Mercer (2017).

References

Alcor. (n.d.). About cryonics. Retrieved February 13, 2018, from http://www.alcor.org/AboutCryonics/index.html
Andersen, A. N., Gianaroli, L., Felberbaum, R., de Mouzon, J., Nygren, K. G., & European IVF-monitoring programme (EIM), European Society of Human Reproduction and Embryology (ESHRE). (2005). Assisted reproductive technology in Europe 2001. Results generated from European registers by ESHRE. *Human Reproduction, 20*(5), 1158–1176. https://doi.org/10.1093/humrep/deh755
Chalmers, D. J. (1997). *The conscious mind: In search of a fundamental theory* (Rev. ed.). New York: Oxford University Press.
De Wolf, A. (2015). Cryonics: Using low temperatures to care for the critically ill. In A. De Wolf & S. W. Bridge (Eds.), *Preserving minds, saving lives: The best cryonics writings from the Alcor Life Extension Foundation* (pp. 18–22). Scottsdale, Arizona: Alcor Life Extension Foundation.
Ettinger, R. C. W. (1962). *The prospect of immortality*. Ann Arbor: Ria University Press.
Horsey, K. (2006, May 29). "Twins" born 16 years apart. *BioNews*. Retrieved from http://www.bionews.org.uk/page_12734.asp
Kass, L. R. (2001). Why we should ban human cloning now. Preventing a brave new world. *New Republic, 224*(21), 30–39.
Koch, C., & Tononi, G. (2008). Can machines be conscious? *IEEE Spectrum, 45*(6), 55–59. https://doi.org/10.1109/MSPEC.2008.4531463
McMahan, J. (1995). The metaphysics of brain death. *Bioethics, 9*(2), 91–126. Retrieved from https://www.ncbi.nlm.nih.gov/pubmed/11653058
Mercer, C. (2017). Resurrection of the body and cryonics. *Religions, 8*(5), 96. https://doi.org/10.3390/rel8050096
Mill, J. S. (2009). *Three essays on religion*. Broadview Press. (Original work published 1874).
Minerva, F., & Sandberg, A. (2015). Cryopreservation of embryos and fetuses as a future option for family planning purposes. *Journal of Evolution and Technology/WTA, 25*, 17–30. Retrieved from http://jetpress.org/v25.1/minerva.htm
Nicolia, A., Manzo, A., Veronesi, F., & Rosellini, D. (2014). An overview of the last 10 years of genetically engineered crop safety research. *Critical Reviews in Biotechnology, 34*(1), 77–88. https://doi.org/10.3109/07388551.2013.823595
Pattinson, S. D., & Caulfield, T. (2004). Variations and voids: The regulation of human cloning around the world. *BMC Medical Ethics, 5*, E9. https://doi.org/10.1186/1472-6939-5-9
Sade, R. M. (2011). Brain death, cardiac death, and the dead donor rule. *Journal of the South Carolina Medical Association, 107*(4), 146–149. Retrieved from https://www.ncbi.nlm.nih.gov/pubmed/22057747

Singer, P., & Wells, D. (1983). In vitro fertilisation: The major issues. *Journal of Medical Ethics*, *9*(4), 192–199. Retrieved from https://www.ncbi.nlm.nih.gov/pubmed/6668584

Topping, A. (2016, November 20). Cryonics debate: "Many scientists are afraid to hurt their careers." *The Guardian*. Retrieved from http://www.theguardian.com/science/2016/nov/20/cryonics-debate-science-freezing-human-bodies

Younge, N., Goldstein, R. F., Bann, C. M., Hintz, S. R., Patel, R. M., Smith, P. B., ... Cotten, C. M. (2017). Survival and neurodevelopmental outcomes among periviable infants. *The New England Journal of Medicine*, *376*(7), 617–628. https://doi.org/10.1056/NEJMoa1605566

CHAPTER 2

Resuming Life

Abstract Most objections to cryonics deal with either the unlikelihood that cryonics will succeed in reviving people or the claim that the enterprise as a whole would be undesirable (whether due to high cost or some potential implications). This chapter starts with an analysis of arguments based on the wastefulness of cryonics, as compared with other costly enterprises, focusing on a comparison between cryonics and various investments that could extend the lifespan of a large number of people for many years. Other common objections to cryonics are based on the assumption that the future will not present socio-economic circumstances that would be favourable to the revival of cryosuspended individuals; hence cryonicists would never be revived by future people. Finally, different possible scenarios in which the cryonicist could be revived are considered, ending with a discussion about the possibility that life for a revived cryonicist would not be good enough to justify their investing in cryonics arrangements today.

Keywords Cryonics • Cryopreservation • Neuropreservation • Medical ethics • Bioethics

Objections to Cryonics

In the previous chapter, we began to draw a conceptual map of the cryonics debate by considering some of the most common arguments voiced against new technologies in general. In this second chapter, we will zoom in on arguments and objections that are specific to cryonics.

The first half of the chapter will explore objections based on the supposed wastefulness of cryonics, as compared with other costly enterprises. One valuable resource that cryonicists would be wasting is their own organs: by choosing to store their bodies in liquid nitrogen after being declared dead, cryonicists would waste valuable organs that could be used to save people who need an organ transplant to survive. Moreover, cryonics is far from cheap, and cryonicists could be wasting significant amounts of money. Instead of spending thousands of dollars on a small chance to save their own life, the cryonicist could, for instance, donate it to a charity that would use it to save several lives.

Both of these objections are based on the idea that if one were faced with the choice of either (A) investing in a small chance of personally living far beyond the average lifespan, or (B) investing in the much higher odds of helping others reach an average lifespan and a decent life quality, it would be morally wrong to choose (B).

In the second half of the chapter, we will explore objections based on potential issues with the revival part of the process, starting with the assumption that future generations will only be interested in reviving the cryopreserved under ideal (and not very probable) socio-economic circumstances. It may be that future resource scarcity will have a negative impact on developing the technology needed to revive the cryopreserved. Conversely, if technology and living standards continue to improve at a steady pace well into the future, generations might even enhance themselves to a point where they are so different from today's *Homo sapiens* as to lose all interest in reviving the cryopreserved.

Moving on, we will discuss the possibility that life for a revived cryonicist would simply not be good enough to justify their investing in cryonics arrangements today. It could be, for instance, that cryopreservation turns out to have a catastrophic impact on the body, making survival after cryonics extremely painful and therefore undesirable. And even if the process itself turned out to be harmless, it may be that old-fashioned humans would struggle with the psychosocial implications of suddenly finding themselves alone in the far future.

We will now look at each of these objections in turn.

Waste of Resources

Waste of Organs for Transplants

Cryonicists are often accused of being selfish, because by choosing to be cryopreserved, they end up "wasting" organs that could be used for transplants. Hospitals are always in need of healthy organs, and harvesting and transplanting them from newly deceased people, rather than locking them down in indefinite cryosuspension, could save several lives per donor. Of course, not all bodies can be used for transplants—there is, after all, no sense in asking a cryonicist who is dying of multiple organ failure to donate their organs—but we can agree that cryonics probably does reduce the total number of potential organs available for transplanting.

One practical solution to this problem would be to preserve one's brain only, whilst leaving the body behind for organ harvesting. This option, called *neuropreservation*, is offered by some cryonics providers (among them Alcor, Oregon Cryonics, and Kriorus). Since, as we have seen in the previous chapter, what is crucial for future revival is that the information stored in the brain remains recoverable, it makes sense to focus on cryopreserving the brain only. Neuropreservation does seem to be the more altruistic option; not only does it not interfere with organ donation, but it is also significantly cheaper than full-body cryonics. Since, as we will see in the next section, one of the common objections to cryonics is that the money spent on cryonics could instead be donated to more effective causes, saving money on cryonics can be considered a less selfish option.

Neuropreservation, in addition to being considerably less expensive than full-body cryopreservation, is also seen as a preferred option for more practical reasons. For instance, in the case of an emergency requiring evacuation of cryonics patients from a facility, neuropatients could be moved faster, more easily, and with less chance of damage than whole-body patients. In the future, it may even become possible to grow an entire body from scratch. If so, it might even be easier to just "create" a new (perhaps even enhanced) body in a lab and connect it to the brain of the neuropatients, rather than modify the body of a patient whose entire body was cryopreserved. The new body could also be made to incorporate features attractive to the owner but not present in the original.

Even if we set aside such speculation, there are good reasons for today's cryonicists to choose neuropreservation over whole-body cryonics. According to the cryonics provider Alcor, new and improved cryonics technologies are often available to neuropatients earlier than to whole-body patients, and it is easier to ensure that cryoprotectants are optimally absorbed by the brain's tissues when only the brain is stored.

However, neuropreservation poses difficulties that can easily be avoided by whole-body preservation. With neuropreservation, the connection between nerves in the brain and the muscles they move is lost. Although it is possible in principle to rebuild these connections, or at least a large part of them, the problem is easily avoided if one chooses whole-body cryonics. Over the past few years, there has been some debate over the feasibility of a brain transplant (or whole-body transplant, depending on one's view) (Telegraph Video, 2016). This kind of surgery has never been attempted on living humans so far, but at least one renowned neurosurgeon has vowed to solve the problem within a few years (Canavero, 2013). Most doctors are very sceptical, however, arguing that the surgery is too complex to become feasible in such a short amount of time. A success in this area would lend more confidence to the idea that a neuropreserved brain could be connected to a new body.

Alternatively, some cryonicists think that it may in the future become possible to upload information stored in the brain to a computer, enabling one to live without the need for a vulnerable biological body.[1] In such a hypothetical future world, people could even choose to upload their consciousness to a virtual universe. Science fiction has produced numerous depictions over the years of what such worlds may look like; one famous recent example is the simulated world of San Junipero from the eponymous episode of the BBC series *Black Mirror*. Inhabitants of this computer-simulated world can live as brain uploads forever, thereby defeating human mortality. Although such an option would almost certainly require far more advanced technology than what is required for transplanting a brain into a new body or reviving an old body, it may, nevertheless, be feasible.

But regardless of the future of virtual reality and the possibility of creating a real San Junipero at some point in the future, it seems that someone opting for neuropreservation could still donate most of their organs, in which case the objection based on organ waste would only apply to whole-body cryonicists. But as we will now see, this does not mean that whole-body cryonics can automatically be cast aside as morally impermissible.

Organ donation is rightly considered a generous choice, but it is not obvious that cryonics should be considered immoral solely on the basis that it would reduce the number of potential organs available for transplants. It seems that the argument's strength is inversely proportional to the degree of confidence one assigns to the possibility that cryonics will work. If cryonics has *zero* chance of working, then of course whole-body cryonics is a *certain* waste of organs. With a non-negligible chance, however, any organs in the cryopreserved body might be used again by the person who originally "owned" them. If one has a high degree of confidence that cryonics will succeed (as a cryonicist might), the perspective shifts. Blaming the cryonicist for not donating their organs would then be similar to blaming a person with cancer for not choosing euthanasia before all their organs are attacked by metastasis and no longer suitable for donation. A person diagnosed with a condition that will almost certainly kill them in little over a year could give up on that one year of life to donate his or her organs and thereby save the life of other people. But even if this might be the most altruistic choice for a person in such a situation, it does not necessarily mean that any other choice would make them blameworthy. Indeed, most people would agree that one extra year of life, especially when one knows it will be their last, is extremely valuable. Sacrificing the last year of one's life in order to donate their organs may be the most generous choice to make, but it is also supererogatory.

Finally, given that only a tiny minority of people currently choose cryonics (and the majority of them choose neuropreservation, anyway) it seems that, at least for now, the impact of cryonics on the availability of transplantable organs is negligible. If one is committed to solving the problem of scarcity of organs for transplantation, attacking cryonics seems to miss the target. There are more effective ways to increase the number of organs for transplant, such as legislation requiring all citizens to be organ donors unless they explicitly opt out. Of course, if the number of people who opted for cryonics drastically increased over time, the impact of cryonics on the availability of organs would be quite significant. Hopefully, by the time cryonics becomes a mainstream option, it will be possible to manufacture organs in vitro from stem cells of the person in need of the transplant, thereby eradicating organ scarcity altogether.

Waste of Money That Could Be Used for Donation to an Effective Charity

When human life is at stake, how should money be spent to realize the greatest good? This is a very touchy subject, as it confronts us with moral paradoxes in some of our most deeply held values.

For example, if one had to choose between one strategy that might save some lives but sacrifice others and another with a different mix of expected lives saved and lost, which should one choose? The choice would be complicated by details such as the expected quality and length of the lives to be saved or extended, versus the deficit for the lives lost. The perceived rightness of each alternative could be heavily affected by such issues as one's religious beliefs (or lack thereof), whether oneself or loved ones are part of one group but not the other, or other values and perspectives one may have or lack.

Imagine a cryonicist who has a high degree of confidence in the likelihood that cryonics will work. To such a person, cryosuspension at clinical death would appear to be a means of powerfully extending one's life in a state of good health. Cryonicists often reject the notion that current limitations on lifespan are anything but a terrible tragedy, and wish for the abolition of practices that sacrifice chances of revival after what is (mistakenly, according to them) considered death, such as burial and cremation. The cryonicists could have a very different perspective on issues of life and death than someone who is not a cryonicist and who does not normally contemplate the possibility or desirability of extending life beyond present natural limits.

With these thoughts as a caveat, we can consider some of the pertinent issues in deciding about cryonics versus donations to charity. First, let us consider the cost of cryonics. A membership at Alcor is about $700 a year; at the Cryonics Institute, it is about $120 per year. Whole-body cryosuspension with Alcor costs about $200,000; neuropreservation is cheaper, at $80,000. The Cryonics Institute only offers whole-body cryosuspension, but with a substantially smaller fee of $28,000. That fee, however, does not include an in-field cryoprotection followed by transportation to their storage facilities (a package they offer for an additional fee of around $90,000), nor does it include an automatic contribution to a trust fund, as with Alcor, so one would probably have to save money in a personal trust fund. Alcor uses about half of its $200,000 whole-body fee to fund a Patient Care Trust meant to ensure the continued operation of their facilities—and hence the safety of their cryopatients—even in case of a severe financial crisis and subsequent lack of funding. Since such problems have caused other cryonics service providers in the past to close down and lose their patients (Perry, 1990/2015), Alcor has tried to develop preventive measures that would avoid such an outcome.

All in all, the total cost of cryonics—ranging between $28,000 and $200,000—is not negligible, but is not prohibitive, either. Indeed, it is accessible even to people with a relatively low income, especially considering that some insurance providers have cryonics packages that cover costs with a monthly fee between $30 and $100, depending (like most insurance prices) on one's age and state of health at the time of enrolment. Still, it is hard to escape the fact that cryonics is, all things considered, expensive. The $200,000 fee required by Alcor for full-body cryosuspension would instead make a generous donation to an effective charity and save the lives of many people.

According to the meta-charity GiveWell, "As of November 2016, the median estimate of our top charities' cost-effectiveness ranged from ~$900 to ~$7,000 per equivalent life saved (a metric we use to compare interventions with different outcomes, such as income improvements and averting a death)" ("Cost-Effectiveness", 2017). The saving of lives often amounts to effective treatments for diseases or parasitic infections that are rampant in some parts of the world, such as malaria and schistosomiasis. For the would-be cryonicist, this means that between 28 and 220 lives could be saved with the same sum of money that is necessary for full-body cryonics at Alcor. The cheapest option, the $28,000 fee of the Cryonics Institute, if invested in donations to effective charities, would potentially save between 4 and 31 people.

So between 4 and 220 lives could be saved if a cryonicist gave up on plans to be cryopreserved and instead donated the same amount of money to effective charities. These considerations could be seen as good reasons to choose charitable giving over cryonics, especially if one is a utilitarian and thinks that there is no particular moral reason that justifies being partial towards one's own life over someone else's. But even if one is not a utilitarian, and considers partiality towards one's own life to be morally permissible, there would still be an unjustifiable discrepancy between the value one can reasonably attribute to their own life and the value they should attribute to the life of other people—especially when we are considering saving as many as 200 lives.

If cryonics has only a small chance of working, the moral imperative of not choosing it would seem to be even stronger. But even if its chances of success were close to 100%, cryonics would arguably remain a choice between saving one life only (oneself) versus saving 4 to 220 lives, albeit not including oneself. There are other considerations, however, cautioning against a summary rejection of cryonics on moral grounds, as we shall now consider.

To begin with, suppose for the sake of argument that we do, in fact, accept that giving to effective charities is a better use of funds than cryonics, and the preferred choice from a moral standpoint. So we should not choose cryonics. But then, on the same grounds, we would have to reject many other conventional yet costly medical interventions aimed at extending the lifespan of severely ill people.

The use of intensive care units (ICUs) for very old people seems an interesting comparison. According to at least one study, the cost of being treated in an ICU is about $5000 per day, admission costs are $32,000, and the average duration of one stay per elderly patient is around one week (Chin-Yee, D'Egidio, Thavorn, Heyland, & Kyeremanteng, 2017).[2] Estimated ICU costs were $49,000 per survivor to discharge (i.e. for a person that left the hospital and did not die in the ICU) and $62,000 per survivor at one year. However, only 33% of these patients survived for more than one year after the ICU treatment. This means that, to some people, the last week of their life cost roughly $67,000.[3] Even those who survive for more than one year after their stay in the ICU are spending a significant sum of money to buy only a few months or years.

This is only one of many examples showing how people are willing to spend considerable amounts of money to extend their lifespan by just a few years. Of course, to say that people do this rather than donate the same amount of money to save the lives of several others is not a good argument for either cryonics or ICUs. But this comparison suggests that cryonics is at least not a markedly more selfish attempt to extend one's life than some widely accepted medical interventions.

The case of Charlie Gard, an infant affected by a rare genetic disorder, was widely debated in 2017. Charlie's health was severely compromised, and the doctors in the UK hospital where he was treated decided to stop life-support treatments and start palliative care. The parents disagreed and asked to try an experimental treatment available only in the United States. While the judges were in the process of determining what was in the child's best interest, Charlie's parents raised £1.3 million in donations so that they could afford taking Charlie to the United States and try the experimental therapy. In the end, they were not given permission to go, and the child died soon thereafter. The debate around Charlie's case touched upon various ethical issues related to autonomy, best interest, and parental rights. However, it was never debated whether it would be moral to try to save the life of one child when between 240 and 1700 people could have been saved if the same amount were donated to effective charities. This case, like many similar ones, shows that there are many instances where people feel it is morally permissible—or, at least, not morally blameworthy—to spend conspicuous sums of money to save one life rather than many.

To say that people often fail to make the best decision does not mean they are absolved of responsibility for the choices they make. We do have a moral obligation to at least consider whether our money is being invested in the most valuable goal according to our values. If we care about saving lives, we should think about how to maximize the number of lives we can save, or, more precisely, how many quality-adjusted life years (QALYs) we can give those who need it most. But it is inconsistent to blame cryonicists for violating this principle in their attempt at life extension while not raising the same objection in cases of similar choices in other contexts.

Deciding to bring a new individual into existence is yet another choice that we often tend to assume is always morally good, or at least morally neutral, and definitely more morally justifiable than cryonics. However, in the United Kingdom, raising a child up to the age of 21 is estimated to cost an average of £229,251 (Bingham, 2015). If this sum were instead donated to a highly effective charity, between 42 and 330 lives of people in developing countries could be saved. This case is particularly interesting because we are not weighing the interests of one person to have their life extended against the interests of several people to have their lives extended or saved. Instead, we are comparing bringing into existence someone who has no interest in existing—non-existent individuals cannot have interests, after all—over saving the life of many existing individuals who have an interest in continuing to exist. Still, most people think it is morally permissible to have children even though doing so would not address any problem of under-population (if anything, quite the contrary), and many existing people already struggle to survive.

One could object that, unlike expensive life-saving treatments and having children, cryonics is an investment with no returns—if one assumes that cryonics has zero chance of working, that is. People who use their money to extend their lives even for just a few months or years usually benefit from the expensive treatment they pay for, which cannot happen for a treatment modality with no chance of success.[4] However, the probability that cryonics will work at some point in the future at least seems greater than zero,[5] so the argument needs adjustment. Cryonics may still have only a small chance of success, unlike other selfish but more conventional investments (medical interventions included), but the difference between no chance and a small chance could be a large one if the gain is potentially very large, as is certainly true with cryonics, where the potential to live hundreds or even thousands of years is at stake. In other words, even if the chance that cryonics will succeed is very low, its expected utility could still be very high.

We need to remember that the long-term goal of cryonics is to transport individuals to a future time when technology will be able to extend human lifespan well beyond current limits, possibly even indefinitely. So a person embarking on a cryonics project is not gambling a large sum of money on the chance to live, say, only another decade or two, but rather on the chance to live for centuries, millennia, or maybe much longer still. Unlike the case of the ICU or other expensive conventional treatments aimed at extending the lifespan of the severely ill, cryonics has the potential to add very many years of life—QALYs, to be precise—to the person involved. Thus, if cryonics works, it will be arguably one of the most cost-effective investments one could make in terms of QALYs gained.

If we assume that $200,000 (the cost of a high-end cryonics plan) could save 220 lives if donated to an extremely effective charity, and that each of these lives would be extended by 50 QALYs on average, we would pay $200,000 to gain 1100 QALYs. This would undoubtedly be a huge gain; but it is important to note that cryonics aims at achieving even more than this for *each individual*, possibly by an infinite amount (since their goal is to reach immortality).

Now, let us assume, for the sake of argument, that cryonics will only allow people to add 200 years of life after revival in the future. Would it be morally justifiable to choose to add these years to the life of one single cryopreserved individual, instead of using the same resources to extend the lives of 200 people by 50 years each?

As with the other cases, such a biased investment might not be morally justifiable, but it seems that it would at least be more justifiable than paying for intensive care (regardless of the age of the patient, because our current lifespan is far less than 200), or having a child (since individuals who do not yet exist do not have an interest in existing), or paying for an expensive experimental treatment to treat a rare disorder (which would usually extend the lives of a very few people, and only by a few years).

Besides, using cryonics as the only term of comparison for selfish choices does not seem to be reasonable. People spend a lot of money—indeed, usually more than $200,000 over a lifetime—on goods that are not even necessary to their survival: big houses, fancy cars, fashionable clothes and jewellery, digital devices and gadgets, and trips to exotic destinations. One could argue that cryonics is a poorer choice than buying a fancy car because it has a very small chance of succeeding, whereas the fancy car will certainly give at least some joy to its owner, however fleeting it may be. But even though cryonics may have only a small chance of working, its goal is arguably less trivial and superfluous than driving a Ferrari or wearing a pair of Louboutin shoes.

Some people claim that, once the average life expectancy is reached, cryonics or any other treatment aimed at adding years is superfluous, trivial, and selfish. But there are at least two further considerations to take into account: first, not all the people who opt for cryonics have reached the average life expectancy to begin with—indeed, some of them are only children—and, second, the average life expectancy of human beings is not fixed, but has increased slowly but steadily over centuries.

One could then reply that cryonics might be morally permissible only if a person "dies" at a young age, when their full potential has not been reached. But this assumes that the ideal amount of time needed for a person to fulfil their potential corresponds roughly to the average human lifespan. If this were the case, it would be quite a coincidence indeed. It is instead far more likely that humans tend to plan their lives and set their goals within the realistic and seemingly inescapable confines of their life expectancy.

For instance, most people have children before the age of 40 because it is very difficult for women to conceive after reaching that age. Most people argue that this is appropriate, since a person over 40 is too old to look after a child anyway. Yet, after in vitro fertilization (IVF) was introduced, women started having children at an older age, without any noticeable negative effect on either the child or the women themselves. So it seems that people start to plan their lives differently as technology finds ways to stretch the natural constraints imposed by human physiology. Many women now plan to have children in their 40s, and to prioritize their career or romantic life until they feel ready to reproduce. Similarly, if lifespans were stretched to 500 years with the possibility of maintaining the bodily and cognitive capacities of a 30-year-old, people would probably make very different plans, and would have different aspirations and understandings of what a life lived to its full potential ought to look like. One might postpone having children until their second century or beyond, meanwhile becoming expert in many different fields, travelling the world, and developing different skills. To someone with a quincentennial life expectancy, dying at age 300—without having seen the whole world, learned to play every classical instrument, mastered at least 40 languages, and having met one's great-great-great-grandchildren—would seem both limited and tragically premature.

It seems, then, that such arguments against cryonics, on the basis that it would be selfish for people over 80 to live longer, presuppose a fallacious view that the current average life expectancy just so happens to be the optimal amount of time required to live a full life. Moreover, this argument fails to take account of individual circumstances: to a woman living in a country where women are considered private property and are not allowed to study, work, or choose their partners, one could argue that her life would not fulfil its potential even if she lived for 100 years. In contrast, some people have very active, interesting lives that allow them to feel satisfied and accomplished by the time they are in their 30s. Of course, the more time a person has to live, the more likely he or she is to have the time to reach his or her goals and feel fulfilled; but the opportunities that come along during one's life can be very different. It is also worth noting that one's interests and goals may well expand over time, so that, as the end of life draws near, what may at first have seemed a sufficiently fulfilled life may not seem that way any longer.

To sum up, cryonics can be considered a selfish investment with possibly a very low potential to succeed. However, selfish investments are very common, yet non-cryonicists are not stigmatized as much as cryonicists are for choosing their own life extension over that of other people. It is true that, unlike other selfish choices, we are currently unable to assess whether cryonics will be successful. This is to be expected; uncertainty is, in many ways, the hallmark of progress. Most of the medical treatments and technological devices we now use on a daily basis were inconceivable a few centuries ago. Being able to conceive of something often catalyses a series of events that eventually moves the concept from the world of ideas to hard reality. We do not yet know if cryonics and immortality will eventually become reality, but we know for sure that not trying at all will preclude it from happening altogether. Since the potential gain is high, it seems that cryonics and life extension should not be brushed aside without having first been carefully considered.

INDIFFERENCE OF THE FUTURE

The success of cryonics depends to a great extent on the motivation of future people to revive cryopreserved patients and to develop the necessary technological tools to do so. In our world, intergenerational obligations tend to become weaker over time. For instance, one might feel a strong obligation to respect one's parents' or grandparents' wish not to have the family house sold even after they passed away, but this kind of obligation usually weakens after a few generations. The possibility that cryonics will only work in the distant future casts serious doubt on the likelihood that future people will have any interest in reviving people to whom they have no obligation or emotional attachment.[6] This objection can be subdivided into a number of different reasons why future generations may not be willing to revive cryopreserved individuals, which we now consider.

No Interest in Spending Resources on Reviving the Cryopreserved

One reason why future people might not want to revive the cryopreserved could be a lack of resources. Suppose, for the sake of argument, that resource scarcity remains a problem well into the future. People continue to suffer from lack of food, clean water, and space in which to live. Most do not welcome the idea of sharing their limited resources with people who had their share of goods and services when they were alive. In this future world, we could justifiably expect that individuals would prioritize saving the lives of their contemporaries over reanimating cryonicists from the (possibly quite distant) past. If such a scenario were to last forever, the cryopreserved would not have a shot at a new life. However, we have reason to believe that resource scarcity of this sort would not necessarily last forever; as history shows, times of poverty and scarcity have tended to alternate with times of plenty in Malthusian cycles of boom and bust. Hence, there is a chance that our future world of resource scarcity would eventually transition into one of plenty, in which we discover new ways to extract resources, for instance, through drastic improvements in terrestrial recycling, solar power generation, or asteroid mining.

In the era of abundance that would follow such developments, future citizens would likely be more inclined to revive the cryopreserved. It is only in a scenario of increasing poverty without a subsequent upswing that revival would become quite unlikely—but in that case, as we will see later in this chapter, it would be in the best interest of the cryopreserved not to be revived into a world where the quality of life is very low.

Even though it is not possible to predict the future by looking at the past, we should at least find some encouragement in an unprecedented trend towards greater prosperity we have witnessed during the past three centuries. We cannot tell with absolute certainty that future people will be wealthier than us, but the fact that we are definitely wealthier than our ancestors suggests that an optimistic attitude towards the future of cryonics is at least not entirely unfounded rational. Moreover, since future prosperity is crucial to the success of cryonics revival, a cryonicist would actually have a strong incentive to help improve the future. Many of our civilization's most urgent problems, from climate change to antibiotic resistance, stem from the fact that we find it notoriously difficult to truly care about future generations, yet all too easy to care about ourselves. The prospect of actually having a personal stake in the far future could, therefore, help bridge this empathy gap between current and future generations.[7]

No Interest in Developing Cryonics Technology

It is also possible that future people will discover alternate ways of indefinitely extending human lifespan, leaving them with little or no interest in improving cryonics technology. Life extension through cell rejuvenation, for instance, holds promise for extending life well beyond the current lifespan, possibly to an indefinite length (as we will see in the next chapter). Cryonicists themselves are indeed relying on this kind of technology someday being developed, but since it is not yet available, they need cryonics in order to "live long enough to live forever".

It is generally assumed that cryonics and life-extension technologies will develop more or less at the same pace. In such a scenario, when rejuvenation and life-extension therapies have become available, cryonics technology will also have developed sufficiently to allow for the revival of the cryopreserved and for the use of rejuvenating and life-extension technologies. However, it is possible that rejuvenation and life-extension technologies will develop much faster than cryonics revival capabilities so that interest in cryonics revival will fall and research will stagnate. In such a scenario, future generations will not have good reasons to invest resources in cryonics because it would be useful only to those few hundreds (or maybe few thousands) of people who have been cryopreserved in the past. However, insofar as life-extension technologies will keep dealing with biological bodies, it is unlikely that interest in cryonics would be lost. Biological bodies like ours will most likely remain vulnerable to accidents and diseases, so cryonics would probably be used in the future to give doctors and scientists enough time to find adequate therapies. Moreover, most revival technologies, such as medical nanotechnology and brain emulation, will likely be developed (or at least pursued) by future generations regardless of their interest in cryonics, since there are other incentives to developing them. So there is a chance that the technology required to revive and heal cryonicists will be developed for reasons that have little to do with cryonics, before potentially being applied to cryonics revival.

No Interest in "Homo sapiens"

In another, perhaps more speculative future scenario, *Homo sapiens* as we know them could disappear altogether and be replaced by more advanced "posthumans". Such individuals might have little interest in bringing back to life the cryopreserved members of a different, less evolved species. They might view humans much in the same way as we view monkeys, lizards, or even worms. Our species may also have major deficits in other ways, such as being unable to breathe the earth's atmosphere as it has become in the distant future. The cryorevived may thus have to live in special, expensive facilities, or perish. Meanwhile, human intelligence level could be too low to function independently in the posthuman world, so that the cryorevived would have to receive constant assistance in order to survive. In any case, if future posthumans come to think of humans as worm-like, they might have little interest in reviving them—except, possibly, for a few representative specimens, who would live more or less as laboratory or zoo animals. The rest might simply be killed or otherwise discarded when it is determined that they are of no further interest.

But even in a less dark scenario where future individuals are not hostile towards the cryopreserved, we have to acknowledge that the capacity to socialize with future posthumans would depend, to a large extent, on the differences between the cryorevived and the future people.[8] If we were to be seen in the same way we ourselves view dogs, it might actually be possible for some socializing to occur. But the disparity with our posthuman stewards could be much greater still, comparable to that between ourselves and, say, mosquitoes—which would make it quite difficult to bond and develop meaningful relationships with our superiors. It is also possible that they would try to enhance us and make it easier to fit within their societies, but it is hard to predict whether they would have any interest in doing so and whether they could actually succeed. And even if the posthumans were extremely benevolent towards the cryorevived, they might be unable to understand their need for friendship and love, or what a meaningful life is to them—just as we are too different from mosquitoes to make anything but highly speculative assumptions about how to optimize their well-being (assuming we were interested in doing so). One might argue that this comparison is not accurate because future persons would probably have technology allowing them to make accurate assumptions about our well-being, for instance, through realistic simulations of human brains. If so, it seems like the best hope for cryonicists to lean on in this context.

In another possible dystopian scenario, future persons could think of cryopreserved people as slaves and bring them back only to use them for scientific experiments or to perform jobs that the posthumans would find too tiring or humiliating. Humanity has a long history of underestimating the moral value of species or races that are considered "other" than the dominant one. We cannot be sure that posthumans would not make the same mistake.

Cryonicists seem to think that, regardless of these possible dystopian scenarios, the community growing around cryonics would provide a certain safety. On their website, Alcor explains that they constitute a community that includes both living and cryopreserved members. The healthy and living cryonicists have close ties with the cryopreserved ones and work to make the revival possible. Once some people are revived, there will be more members seeking ways to revive the ones who are still cryopreserved, and to whom they feel connected, perhaps in virtue of relationships they had during their life before cryopreservation. The success of cryonics is hence largely dependent on the strong sense of community of the cryonics group: "The plan is not for 'them' to revive us. The plan is that we, the Alcor community, will revive ourselves" (Alcor, n.d.).

Desirability of Being Revived in the Future

In an optimistic future scenario where future humans or posthumans revive the cryopreserved, there could still be circumstances that would make the life of the cryorevived very painful and, in some cases, not worth living at all.

For instance, as mentioned above, it could be that revived cryonicists would only be able to survive if housed in special facilities resembling hospitals. Perhaps to forestall some intractable psychological trauma, they would have to be drugged to a permanently numbed state that made them incapable of appreciating that they actually "travelled through time" and were alive again. Or maybe their bodies, affected by the time spent in liquid nitrogen in ways we cannot foresee, would suffer from unbearable pain that could not be remedied in any way.

More generally, it is possible that the future world would be an unpleasant and dangerous place in which to live. Wars, climate change, or political and social events might turn the planet into something no human would care much to inhabit, notwithstanding humans' capacity for adapting to difficult circumstances. Under such circumstances, the cryorevived might choose to be euthanized, and cryonics would then have been a failure and a waste of resources despite being technically successful.

It is not clear, of course, whether and how any of the above scenarios would happen, but we cannot rule them out. Depending on how unpleasant life in the future world would be or would be perceived to be, the cryonicist would need to decide whether they want to keep living or just end their life once and for all. In such a gloomy scenario, despite cryonics having succeeded in transporting cryonicists to a future world, we would have to agree that it would still have been a failure, as it would not have enabled cryonicists to add any valuable time to their life. And even though some cryonicists might consider having at least a peek into the future as a better outcome than not seeing the future at all, they would still probably agree that things did not exactly go according to plan.

Trouble Adapting Even to an Objectively Better World

It is also possible that the world of the future will be far better than the one in which we currently live. Indeed, certain historical trends seem to indicate that we are headed in this direction (Pinker, 2011). At the very least, it seems that our lives are, on average, much more comfortable and safe than the lives of our ancestors. If the quality of life is going to increase steadily over time, the future might be wondrous indeed.

We might succeed in developing rejuvenating therapies that extend our lives for millennia and more. We might develop enhancement technologies that make us smarter, faster, and better at doing just about anything. Robots might replace humans at most jobs. Future people could then go on a permanent holiday if they wished, or have the time and resources to take up any job they fancy, and at their own pace. That waking up in such a world would be a very positive experience seems like a no-brainer. Indeed, curiosity about the future, together with optimism about the future, is a common reason why people make arrangements for cryonics.

But even in an idyllic scenario, life might still be difficult for the cryorevived. The environmental, cultural, and political changes that occurred over centuries or millennia might cause them to feel extremely out of place, confused, and uncomfortable. Old people today already struggle to fit in a world where the digital revolution has radically changed the way people communicate, work, and spend their leisure time; yet this is nothing compared to the stupendous changes a revived cryonicist might encounter. A sense of alienation could pervade their existence, and navigating such a different cultural and social landscape could be daunting, if not downright traumatizing. Cryorevived people could, of course, find comfort in each other's company, but if the number of cryorevived were tiny, alienation might still override.

Cryonics optimists usually meet this objection by comparing their planned future experience with a journey in today's world or in times past. Waking up after being cryopreserved, they say, could be similar to moving to a new country that is far from the sights, sounds, and people with whom one is familiar. Indeed, it might not be that different from moving from Europe to Australia more than a century ago, where lack of internet communication and cheap airline flights made the journey much more arduous and stressful than it would be now. The first Europeans who moved to Australia had to adapt to a different climate and environment and create new social and family connections, and the cryorevived would similarly need to adjust to—and, to some extent, shape—a new world for themselves. But the eighteenth-century colonists eventually found a way to deal with their new life, and it is possible that the cryorevived would also end up building new relationships and ultimately enjoy their new existence.

Besides, one possible solution to the adaptation problem would be signing up for cryonics with at least another significant person, be it a partner, a family member, or a good friend. Some features of being alive again in a world that is radically different might still be distressing, but sharing the experience with a loved one would make the whole less of a burden and more of a gift.

But even without a significant other to share the experience, the future could still be well worth it. The lonely cryonicist could accomplish much in a few centuries to make life worthwhile and eliminate negatives. One possibility might just be to gradually erase older, unpleasant memories, or even good memories which still caused distress for other reasons. Some memories of life before cryosuspension—say, of loved ones who did not get cryopreserved and are now gone—might cause so much distress that it would be preferable to forget them. People in their 80s and 90s today often have few memories of their lives as teenagers, so the sort of partial forgetting we are talking about does have its precedents in our world.

Overall, the family and friends whom a cryonicist would miss upon revival would probably seem less central to their lives as they became several centuries old, and the pain caused by the absence of the departed would be replaced by the joy of sharing a new life with others.

It could also be argued that the destruction of memories is bad and should be avoided if possible, as these are, in effect, historical records which ought to be preserved like other records of the vanished past. The cryonicist who is unhappy because of remembering too much (including memories of departed loved ones who were not cryopreserved) could elect to have certain memories blocked or made inaccessible, but not erased.[9] One variant of this would be to copy these memories and store them in a safe place before they are erased from the owner's brain. That way, the memories might be restored, when the cryonicist has lived many long and happy centuries and has matured to the point that remembrance will no longer cause distress. The maturation process itself could involve an increase in intelligence, understanding, and wisdom. A world of posthumans who neither age nor die could offer many opportunities for meaningful and rewarding activities, from solving mysteries of the cosmos to sampling the vast smorgasbord of conscious experience.

Some of the possible future scenarios we explored are far from encouraging and might even be severe enough to dissuade some people from choosing cryonics. But there is as yet no proof that such scenarios are more accurate, or more likely to become a reality, than the more optimistic ones. To some people, especially those who attribute great value to living longer and living in the future, it will be worthwhile to give cryonics a try regardless of any possible mishaps that could occur. They are not dissuaded by the thought that things could go awry.

Uncertainty, after all, is a hallmark of human existence. To a certain extent, we are gambling on the future in all our investments: we buy houses in places that look breath-taking, but that in a few years' time could appear completely different due to wars, earthquakes, or other unpredictable events. We choose to share our life with someone who seems to embody everything we could desire in a human being, but who, for all we know, might turn out to be a very skilled sociopath. We invest time, money, and other resources in projects that never materialize or turn out far worse than we expected.

Cryonics is not so different from all these other long-term investments. Luck may play a part in whether the enterprise is successful, yet hard work still improves one's chances and makes one "luckier". As ever, nothing ventured, nothing gained.

Notes

1. For a discussion of future potential brain uploading technologies, see Chalmers (2010).
2. The study covers the period 2009–2013. Cost figures in the article are given in Canadian dollars but exchange rates in US dollars hovered around 1.0 during this period, so parity between the two currencies is assumed here. Figures given in the article are rounded to the nearest thousand.
3. (7 × 5000) + 32,000.
4. Some of the dues and funding for cryonics goes to organ preservation and ischemia research, so even if cryonics were to fail, not all of the money spent on cryonics-related research would result in a waste of money.
5. For an interesting discussion on the online forum Less Wrong regarding the probability of cryonics success, see Kaufman (2011).
6. It is worth noting that there are some organizations, such as The Venturists, that address such potential issues by offering support to people who might be revived in the future.
7. Thanks to Adrian Rorheim for suggesting this implication.
8. For an illustrative and highly entertaining example of what such an encounter might look like, see Wahls (2017).
9. Thanks to Mike Perry for highlighting this issue.

References

Alcor. (n.d.). Cryonics FAQ. Retrieved February 13, 2018, from http://alcor.org/FAQs/faq01.html

Bingham, J. (2015, January 22). Average cost of raising a child in UK £230,000. *The Daily Telegraph*. Retrieved from http://www.telegraph.co.uk/news/uknews/11360819/Average-cost-of-raising-a-child-in-UK-230000.html

Canavero, S. (2013). HEAVEN: The head anastomosis venture Project outline for the first human head transplantation with spinal linkage (GEMINI). *Surgical Neurology International*, 4(Suppl. 1), S335–S342. https://doi.org/10.4103/2152-7806.113444

Chalmers, D. (2010). The singularity: A philosophical analysis. *Journal of Consciousness Studies*, 17(9–1), 7–65. Retrieved from http://www.ingentaconnect.com/content/imp/jcs/2010/00000017/F0020009/art00001

Chin-Yee, N., D'Egidio, G., Thavorn, K., Heyland, D., & Kyeremanteng, K. (2017). Cost analysis of the very elderly admitted to intensive care units. *Critical Care/The Society of Critical Care Medicine*, 21(1), 109. https://doi.org/10.1186/s13054-017-1689-y

Cost-Effectiveness. (2017, November). Retrieved January 10, 2018, from https://www.givewell.org/how-we-work/our-criteria/cost-effectiveness

Kaufman, J. (2011, September 25). How likely is cryonics to work? Retrieved February 15, 2018, from http://lesswrong.com/lw/7sj/how_likely_is_cryonics_to_work/

Perry, M. (2015). Suspension failures: Lessons from the early years. In A. De Wolf & S. W. Bridge (Eds.), *Preserving minds, saving lives: The best cryonics writings from the Alcor Life Extension Foundation*. Alcor Life Extension Foundation. Retrieved from https://market.android.com/details?id=book-6QgvjgEACAAJ (Original work published 1990).

Pinker, S. (2011). *The better angels of our nature: Why violence has declined*. Viking. Retrieved from http://www.ceeol.com/content-files/document-263597.pdf

Telegraph Video. (2016, September 20). Russian man set for world's first head transplant. *The Daily Telegraph*. Retrieved from http://www.telegraph.co.uk/news/2016/09/20/russian-man-set-for-worlds-first-head-transplant/

Wahls, J. (2017, June 5). Utopia, LOL? Retrieved January 11, 2018, from http://strangehorizons.com/fiction/utopia-lol/

PART II

Cryonics as a Step Towards Immortality

Introduction

Most people who make arrangements for cryonics hope to "die" at an old age and be revived in a future where medicine is so advanced that the disease that caused them to die can be fixed. In this second part of the book, we will consider the ethical issues related to cryonics understood as a step towards indefinite life extension and possibly immortality.

Some scientists think that rejuvenation technologies could be developed in a not-too-distant future, at which point a person's entire body could be kept young indefinitely (de Grey & Rae, 2007). While it is not clear yet how such treatments would work in practice, the general idea is that one could regularly rejuvenate all of one's body at a cellular level—from neurons to liver cells—so as to remain highly functional and healthy for an indefinite time.

The reason why cryonics is expected to be paired with rejuvenating treatments is that cryonics by itself can only provide life *suspension*, as it were, but not life *extension*. That is, it can only temporarily suspend the processes that normally cause ageing, and these processes resume immediately upon revival unless additional steps are taken to halt them further. Hence, especially if one undergoes cryopreservation at an advanced age, one merely postpones the last few years of one's experienced life, without actually extending the total amount of time one has spent alive in the biographical sense of being alive.

Let us imagine an 80-year-old patient who, after having endured Parkinson's disease for a decade, decides to get cryopreserved in the hope that he or she may someday be revived and restored to health. One day, a century or so later, a cure is discovered that not only prevents Parkinson's, but also reverses any damage already caused by the disease. Our cryopreserved patient is revived, cured, and restored to the level of physical health he or she enjoyed before the onset of the disease. However, he or she remains 80 years old and thus near the end of life.

Now, let us imagine that the aforementioned cure for Parkinson's is discovered alongside hundreds of equally groundbreaking developments during a new medical revolution. Among other milestones in this future society, rejuvenation technologies have also become reasonably cheap, effective, and safe. Just as our newly revived cryonicist would have reasons to be cured of Parkinson's, he or she would also have reasons to restore every other part of his or her health to the levels they were at before the effects of ageing set in—a biological age of around 40, say. In general, it would seem that the best option for the newly revived would be to undergo rejuvenation therapy so as to turn their biological clocks back however far they want and then regularly undergo rejuvenation in order to remain within a certain age range. Indeed, it seems that rejuvenation would be one's only real hope of avoiding all the degenerative effects of ageing that eventually lead to death. Although some people die of cancer, heart attacks, or strokes at a young age, it is vastly more common for such conditions to befall the elderly. Even the healthiest of old people eventually die because of some unavoidable structural failure, accumulated damage, or both, within their bodies. Even though scientists do not agree about the upper limit of human lifespan (Dong, Milholland, & Vijg, 2016; Rozing, Kirkwood, & Westendorp, 2017), it appears unlikely for anyone to live beyond 150 years without rejuvenating treatments. Hence, it is very plausible that in a hypothetical future where life extension is available, affordable, and routinely used to prevent the development of age-related pathological conditions, ageing would no longer be a matter of fate. It would be a matter of choice.

If ageing were a matter of choice, it is unclear whether one would have any reason to stop undergoing rejuvenation treatments after only a few cycles. After all, if we were now offered a pill that would magically make us five years younger, would not we all be tempted to take it? After how many pills would we decide that it is time to restart ageing at a natural pace and, eventually, die?

Admittedly, though, answering these questions is not as easy as it might seem. This is partly because there are many aspects of immortality that ought to be considered before enthusiastically committing to the project, but also because we humans seem to have a rather ambiguous relationship with death.

At first glance, it seems we fear death above anything. We go to great lengths to avoid dangers and live longer, and to find remedies to any accidents and diseases that might cut our lives short. Even though we place an especially high value on our own life and on the life of the people we love, we still regard almost any death as a tragic event, especially when the victim is young and is being deprived of many years of life. As a society, we ask medicine and technology to buy us more time to do the things we like, to be with the people we love, to explore the world and learn how it works. We want more time to enjoy all the pleasures of life, big and small—from drinking a glass of fresh water, to falling in love, to learning about a complex topic.

If we asked anyone whether they'd prefer to either add or subtract a given number of decent quality years to their life, we can be quite confident that almost everyone would choose to have years added rather than subtracted. Nobody wants to die—or at least, nobody wants to die *right now*, or tomorrow, or over the next few weeks (unless their life entails unbearable suffering, that is). It is not a coincidence that the majority of religions tend to feature one or more immortal deities who promise everlasting life to their adepts: we want to be reassured that once our mortal life is over, a new one will start.

Also, the quest for immortality is perceived by some as the ultimate act of hubris, a selfish and irresponsible attempt to become godlike (Sutton, 2015).

Yet, it is not clear that everyone or even the majority of people, even if they want their current lifespan to be prolonged, would want to be immortal. At least some people would probably say that at some point in the future, they will want to die.

So, on the one hand, death is perceived as the ultimate tragedy, something to avoid at all costs; but, on the other hand, the desire to extend life indefinitely—to become, in a word, immortal—might not be equally widespread.

It is possible that aversion to immortality is, at least in part, explained by aversion to some features commonly associated to immortality, but which are not necessary aspects thereof. Immortality is simply the fact of never dying; this fact could have certain implications that are necessary, such as the fact that one will have an infinite number of experiences, and implications that are contingent and which therefore could be avoided by changing the circumstances in which people live forever. For instance, it is possible that some people do not want to be immortal because they do not want to suffer from the aches and pains that are associated with growing extremely old; however, rejuvenating treatments could eliminate this side effect of immortality. Others may argue that in a world where no one dies, there would be serious problems of overpopulation. But overpopulation is not a necessary consequence of immortality (e.g. we could start colonizing other planets). So some objections to immortality are actually not in-principle objections, but objections to something else. If we are to object to something, then we must consider that thing *in and of itself*, along with its *necessary* (but not the contingent) implications. And if we want to object to some aspect that we associate to that something, we need to consider how bad it is relative to our other alternatives, and how we might otherwise achieve our desired result with fewer bad associated effects.

Hence, in the following two chapters, we will try to understand if there are convincing arguments against immortality in and of itself. In order to do this, we will consider, in Chap. 3, the alternative to immortality—death—and whether, and in what ways, and to what extent, it is a bad thing. This discussion will serve as a premise for a discussion, in Chap. 4, of whether immortality is not only good or bad in itself, but actually better or worse than death.

REFERENCES

de Grey, A., & Rae, M. (2007). *Ending aging: The rejuvenation breakthroughs that could reverse human aging in our lifetime*. St. Martin's Press. Retrieved from https://market.android.com/details?id=book-vlBAKAESSg4C

Dong, X., Milholland, B., & Vijg, J. (2016). Evidence for a limit to human lifespan. *Nature, 538*(7624), 257–259. https://doi.org/10.1038/nature19793

Rozing, M. P., Kirkwood, T. B. L., & Westendorp, R. G. J. (2017). Is there evidence for a limit to human lifespan? *Nature, 546*(7660), E11–E12. https://doi.org/10.1038/nature22788

Sutton, A. (2015). Transhumanism: A new kind of Promethean Hubris. *The New Bioethics: A Multidisciplinary Journal of Biotechnology and the Body, 21*(2), 117–127. Retrieved from https://www.ncbi.nlm.nih.gov/pubmed/27124960

CHAPTER 3

The Death Conundrum

Abstract Cryonics is considered a key step towards indefinite life extension. But are there good reasons to extend the human lifespan beyond its current limit? One possible reason is that death is bad, and since death is bad, we should avoid dying by staying alive indefinitely. In this chapter, possible explanations for why death might be bad are examined. In particular, two accounts of the badness of death are considered in detail: death as deprivation of a future life and death as frustration of desires and preferences. These two accounts of the badness of death provide good reasons for considering death bad under certain circumstances (although not under all circumstances, such as when a life entails unbearable pain), yet leave open the question of whether immortality might be an even worse alternative.

Keywords Death • Indefinite life extension • Deprivation
• Frustration of desires • Immortality

Is Death Bad?

One of the recurring themes in ancient Greek literature is the badness of life and the desirability of death. For instance, in Sophocles' tragedy *Oedipus at Colonus*, the chorus says "To never be born/Is the greatest fortune of all/But once we are born/It is best to return whence we came" (Sophocles, v. 1225–28, auth. trans.). A similar sentiment is expressed by

Queen Hecuba in the tragedy *The Trojan Women*, written by Euripides, in which she says: "Do not think that anyone among the most fortunate people is happy/before they are dead" (Euripides, v. 500–10, auth. trans.). And in the *Histories*, Herodotus first told the tale of the brothers Kleobis and Biton, introduced as the happiest people in the world by the Athenian statesman Solon (Herodotus, 1:61, auth. trans.). According to the legend, their mother had prayed the goddess Hera to reward her sons for their extraordinary devotion and kindness. Hera agreed to reward the two siblings and, as a sign of her appreciation, she let them die quietly a few hours later. These examples are only some of the many tokens of the attitude towards life and death that characterized the ancient Greeks of the fifth century BC. To them, life was considered necessarily painful, and death was greeted as freedom from suffering. This Weltanschauung might seem incommensurably distant from that of the contemporary Western world, characterized by a view that life is good, and by the desire to extend it through all the available antidotes to life-shortening threats.[1] But even in Modern Western societies, although death is generally perceived as bad, not all deaths are perceived as *equally* bad. For instance, the death of very old people does not seem to elicit the same kind of reaction that the death of a child elicits. The (perceived) badness of death seems to steadily drop after one has started approaching the currently average lifespan, to the point that the death of a centenarian is perceived as peaceful as the disappearing of a snowman at the touch of the first rays of sunlight.

Death as Transition to Nonexistence

In philosophical terms, death can be defined as the "unequivocal and permanent end of our existence" (Nagel, 1970). At first glance, this description might seem to illustrate the obvious badness of death—after all, how could the unequivocal and permanent end of existence be good?—but it is, in fact, a value-neutral description of fact. The fact that nonexistence is bad is not as obvious or uncontroversial as we might think. For instance, we saw that in the Greek tragedy tradition, death was not considered bad, but as a desirable way out of the inescapable suffering of life; and nonexistence, be it prenatal or postmortem, was considered better than existence.

So in order to understand whether death qua nonexistence is good or bad, we need to understand if there is something good or bad about nonexistence: if death always entails nonexistence, and nonexistence is always good (or bad), then death is good (or bad).

Around a century after the tragedians Aeschylus, Euripides, and Sophocles lived in Athens, the philosopher Epicurus started what we now know as the Epicurean philosophy. Epicurus argued that death is not bad and should not be feared, but for crucially different reasons than the ones invoked by the tragedians. According to Epicurus in his *Letter to Menoeceus*, death should not be feared because once we are dead, we cannot experience anything:

> Why should I fear death? If I am, then death is not. If Death is, then I am not. Why should I fear that which can only exist when I do not? (Epicurus, v. 124–127, auth. trans. 2017)

Along the same lines, Lucretius, an Epicurean philosopher, argued in *De Rerum Natura* that death is mere nonexistence, and just like we do not worry about the nonexistence that preceded our birth, we should not fear the nonexistence that will follow our death (Lucretius, v. 830–839, auth. trans.). So, in the Epicurean tradition, the ontological equivalence between the nonexistence of the unborn and the nonexistence of the dead is the reason why death is not bad, and the fact a person cannot be harmed once he or she ceases to exist is the reason why death is not to be feared. Although the Epicurean approach has a powerful consolatory message for mortal beings like us, it has implications that are at odds with some of our basic intuitions, including the commonly shared view that killing is (at least prima facie) morally impermissible. If the ontological status of the unborn and of the dead is equivalent, it would mean that not procreating and killing are morally equivalent and both morally permissible.

Some Epicureans might be willing to bite the bullet and say that, indeed, there is no harm in killing someone, or that, more modestly, the immorality of killing is not explained by the harm inflicted on the murdered individual. But non-Epicureans disagree with the view that death is not bad by virtue of being mere nonexistence and that prenatal nonexistence and postmortem nonexistence are ontologically symmetrical.

I will now consider two strategies developed to resist the Epicurean argument; both strategies aim at proving (1) the asymmetry between prenatal and postmortem nonexistence, and (2) the badness of death against the Epicurean claim that death is not bad.

A Life Worth Starting and a Life Worth Living

Contemporary antinatalist philosopher David Benatar has argued, not too dissimilarly from writers in the Greek tragedy tradition, that bringing a new person into existence is harmful. Given that life entails suffering, being brought into existence means to be harmed; however, an individual who is not brought into existence cannot have an interest in existing, and therefore is not harmed by not being brought into existence (Benatar, 2008). Unlike the tragedians, though, Benatar does not consider death the second-best option for someone who is already born; and, unlike the Epicureans, he argues that there is an asymmetry between prenatal and postmortem nonexistence. After someone is born, they develop interests that, in order to be fulfilled, require them to stay alive. Once these interests develop, the bar for considering nonexistence (death) as a better option than existence becomes higher than it was before birth. So, if I am wondering whether I should have a child, and I can predict that my future child will experience over his or her life, say, 116 days of suffering, the conclusion I should reach is that conceiving would be immoral, no matter how much pleasure or happiness their life will predictably contain. Meanwhile, no potential child is going to be harmed if I do not conceive them, because one cannot harm someone by not bringing them into existence, given that one cannot harm someone who does not and will never exist.

Now, imagine that I am considering whether I should mercifully kill my neighbour (let us call her Dolores). She has been suffering for 116 days over the past two years already, and it is highly likely that she will suffer just as much over the next two years. According to Benatar (and common sense), killing Dolores is impermissible, because she (most likely) has an interest in continuing to exist. For instance, it might be that, although Dolores has been suffering due to a failed marriage, the death of a friend, and a long period of unemployment, she, nonetheless, has an interest in raising her children, getting her Master's degree, or finishing a painting. Regardless of the specific motivations she might have to continue to live in the face of adversity, killing her would inevitably deprive her of the chance to pursue her interests, and would thus harm her.

One might argue that this hypothetical case is cumbersome, as it postulates an impossible comparison between the entirely hypothetical life of someone who has not been born and the empirically analysable life of someone who already exists. So let us now consider a different example in which we do not compare two different ontological statuses.

Suppose that I want to go on a last-minute holiday in Portugal. Shortly before I make arrangements for the trip, I check the weather forecast and realize that it is going to rain constantly throughout my planned time there. Given this bleak forecast, I decide to pick a different destination where it is not going to rain. Now suppose that, instead of booking last minute, I had booked my flight and accommodation in Portugal six months before the departure day. I have already invested time, money, and energy into organizing my trip, choosing the places to explore, making arrangements with friends travelling with me, and so on. Given this upfront investment, I would probably go through with the trip to Portugal even after finding out that it will be raining throughout my stay. Once I have invested resources in my trip, I have a fairly strong interest in making the most of it and following the plans I have made. So there is an asymmetry between the "good weather threshold" that has to be met in order to *book* my holiday and the "good weather threshold" that has to be met in order to *cancel* my holiday. In the first case, the foreseen rain is a sufficiently good reason for not going to Portugal; in the second case, the rain is not a sufficiently good reason for cancelling my trip to Portugal. This "threshold asymmetry" mirrors the asymmetry between the "good life threshold" required to consider a life worth starting and the "good life threshold" that would have to be met for someone to choose to die.

But one could argue that the asymmetry between these thresholds is better explained by psychological biases like the status quo bias (the tendency, when faced with potential change, to prefer keeping things as they already are) or the sunk cost fallacy (the tendency to keep investing time or resources in a hopeless undertaking: since one has already invested a lot, giving up would feel like a waste) (Kahneman, 2011). So it may be that one prefers to keep living not because they have developed interests or because they have good reasons to do so, but because they rationalize their current circumstances or their preference for continuing what they have started, even though it would be objectively better for them to quit.

It is hard to tell whether someone continuing to live in the face of serious objective difficulties is just being biased, or whether they are actually showing extraordinary resilience. In most cases, it would not make much difference, because the result would be the same: a person wants to keep living, hence we should let them live. However, we want to understand whether there is an asymmetry between prenatal and postmortem nonexistence, so we will now turn to a second argument that attempts to explain how such an asymmetry may arise.

Whose Nonexistence?

One interesting account of prenatal versus postmortem nonexistence is put forth by contemporary philosopher Thomas Nagel, who argues that postmortem nonexistence is the time during which a now deceased person would counterfactually have been alive (Nagel, 1970). There is an identifiable subject, Dolores, whose life we can imagine continuing had death not occurred.

However, if we think of prenatal nonexistence, we cannot think of it as a time during which an identifiable subject would have lived, had they been born earlier. An individual cannot be considered an individual, identifiable subject until they are, at the very least, an embryo (but probably even at a later stage, e.g. when twinning is no longer possible). It is not possible to turn the clock back to a time before a person's embryonic stage and realistically imagine the counterfactual life of that not-yet-conceived someone. For instance, the lay thought experiment *if I were born ten years earlier* is absurd; I am myself because I was conceived at that exact moment and by those exact parents. Any individual conceived by my parents ten years before they conceived me would, in effect, be my older sibling—not a decade-older version of me.[2]

So, as Nagel argues, the difference between prenatal and postmortem nonexistence can be explained by the fact that before a certain point (be it the development of self-consciousness, birth, or conception) there is no subject to whom we can attribute a counterfactual existence. However, we can easily imagine how someone's life would have kept unfolding had they died later.

Nagel refers to the hypothetical case of future people revived after cryonics to prove that nonexistence after "death" is not bad per se, but is bad when not followed by existence. If one *comes back* into existence after cryosuspension, then they can restart the life they had interrupted, in much the same way that people who are in a comatose state for some time can resume their life when they eventually wake up again. As Nagel observes, the inconveniences of not existing for a long time would not outweigh the fortune of continued, albeit interrupted, existence (ibid., p. 77). However, if revival never becomes available, cryosuspension and postmortem nonexistence would turn out to be equivalent.

Thus, what differentiates postmortem nonexistence from its prenatal counterpart is its being ascribable to a precise subject who would still exist if death had not occurred, while what makes postmortem nonexistence different from cryonics is its being irreversible.

Yet even this conception of death—as the irreversible state of nonexistence of an identifiable individual—fails to explain why death should be considered bad. It explains why not-conceiving, cryopreserving, and killing someone are morally different even if they all imply nonexistence, but it still does not add any normative qualification to the description of death as nonexistence. Indeed, a Greek tragedian could still claim that "death is good because it is the irreversible nonexistence of an individual" while a modern novelist could say "death is bad because it is the irreversible nonexistence of an individual". Either claim could be true, but they cannot both be true. So we need to understand whether postmortem irreversible nonexistence entails some further fact that we can consider necessarily good or bad, or at least that would make death necessarily worse or necessarily better than life. In the next paragraph, we will consider a different account of death that might provide us with a better answer to the question about the badness of death.

Death as Deprivation

Nagel argues that death is bad because it deprives us of the capacity to experience anything, let us call this the "Deprivation Theory". According to Nagel's Deprivation Theory, "[i]f we are to make sense of the view that to die is bad, it must be on the ground that life is a good and death is the corresponding deprivation or loss, bad not because of any positive features but because of the desirability of what it removes" (ibid., p. 76). So, even if the sum of the painful experiences in my life had to greatly outweigh the happy ones, death would still be bad: merely experiencing life, irrespective of the content of such experiences, according to Nagel, "is not merely neutral: it is emphatically positive" and "[t]herefore life is worth living even when the bad elements of experience are plentiful, and the good ones too meagre to outweigh the bad ones on their own. The additional positive weight is supplied by experience itself, rather than by any of its consequences" (ibid., p. 76).

There are two problems with Nagel's views expressed in these claims. First, it is difficult to explain how a dead person could suffer from any form of deprivation. Dead people cannot experience anything (as Epicurus also argued). The second problem is that we have fairly compelling evidence that life is not always a net positive, as we can infer from the fact that some people choose to end it.

The Harm of Deprivation

Let us start with the first issue: how can one *be deprived* of something (i.e. the good that life is) when one no longer exists? If someone deprives me of my laptop, I experience the inconvenience of losing my data or having to spend money to buy a new one. But if someone takes my laptop after I am dead, I am no longer a subject who can experience such deprivation.

Nagel argues that there are certain forms of deprivation that can affect someone even if they never experience their effects. Let us consider a variation on the examples originally suggested by Nagel, and imagine that someone surreptitiously gave me an imaginary "regression pill" that caused me to look, behave, experience the world as a five-year-old for the rest of my life (though it does not affect the number of years I am going to live overall). After unknowingly taking the pill, I would lose my capacity to do philosophy, travel by myself, and achieve any of the goals I have set for my future life. I would have different conversations with people around me, I would have different friends and hobbies, and I would not be able to reconnect to my previous self in a meaningful way. My psychology would be altered in such a dramatic way that my family and friends would perceive me as a different person. In sum, although biologically I would still be alive, the *person* that I currently am would no longer exist; at best, just a small part of my original self would continue to exist.

Assuming that most people want to follow a normal developmental trajectory through their life (one that involves growing older, not younger), it would be reasonable to argue that the regression was a bad thing for me. But how so?

One could argue that, since there is obviously nothing *intrinsically* bad in having the psychology of a five-year-old, or in merely being five years old, the pill did not harm me. However, the quality of my life after the regression is not the only element we would need to take into account in order to assess whether my regression was bad or not. Indeed, even though the "old me" no longer exists and thus cannot process what happened, and even though the "new me" is happy, it would still be true that I was deprived of the life I had planned to live. So even if my old self would disappear and would not be able to process such events, and even if my new self were happy, it would be reasonable to say that the pill overall harmed me because it deprived me of the possible, normal, and desired development of my life. So I can be deprived of something even if I no longer exist, because what I am deprived of is the life I would have lived *if* the pill were not given to me. Similarly, if death is bad because it is deprivation of a future life, it can be bad for a person who no longer exists.

The Plausible Counterfactuals

Let us assume that the above arguments are correct: that death is bad for an individual because it deprives them of their counterfactual life and all the potential value therein. But how do we navigate the countless counterfactuals that we can imagine for each event that takes place in this world?

Nagel argues that "[c]ountless possibilities for continued existence are imaginable, and we can clearly conceive of what it would be for him to go on existing indefinitely. (...) Death, no matter how inevitable, is an abrupt cancellation of the indefinitely extensive possible" (ibid., pp. 78, 80).

I want to suggest that, by including into the possibilities that death would prevent all those that are "imaginable", Nagel is overestimating the badness of death. To start with, although countless, some possibilities are precluded at a metaphysical level. I cannot say: had I not taken the regression pill, I would be Bertrand Russell, because A (me) can only be A and not B (Bertrand Russell). I cannot conceive of being myself and Bertrand Russell at the same time, just like I cannot conceive of a square that is also a circle. It would be false to claim that the person who gave me the regression pill deprived me of a future life as Bertrand Russell. That pill deprived me of many things, but being Bertrand Russell was surely not one of them.

Let us focus now on conceivable possibilities. Now, of course the fact that something is conceivable does not make it a plausible counterfactual. For instance, I can imagine a universe wherein humans have wings, and say: if humans had wings, they could enjoy flying. These counterfactual humans are conceivable, but they do not represent a plausible term of comparison for humans. This counterfactual world is so distant from ours that it would be odd to claim that the misfortune of Dolores having to die in her 30s is the misfortune of missing out on a life with wings. Dolores having wings was merely conceivable, but not really possible, whereas her living up to 80 was not merely conceivable, but also possible or even probable (given basic facts of human biology). Her death should rightly be considered worse than her missed opportunity to enjoy winged flight. We may still deem it bad that Dolores never had the joy of flying with her wings, but we would still have to say that her early death was *worse* than her missed opportunity to live her life with wings.

In order to overcome such difficulties, Jeff McMahan has proposed a variation of the Deprivation Theory formulated by Nagel: the Revised Possible Goods. According to McMahan, "the relevant alternative to death for purposes of comparison is not continuing to live indefinitely, or forever, but living on for a limited period of time and then dying of some other cause". According to this account, the badness of death can be measured "in terms of the quantity and quality of life that the victim would have enjoyed had he not died when and how he did" (McMahan, 1988).

So, just like the badness of the regression pill can be measured in terms of the quality of the life I would have enjoyed had I not regressed to the age of five, the badness of death should be measured against the quantity and quality of life one would have had, had they not died.

McMahan's Revised Possible Goods Theory narrows down the range of counterfactuals against which it is reasonable to assess the badness of death to those that are not only possible, but indeed plausible, that is, that are consistent with living for a limited period of time under normal circumstances. It also explains why, given the current human lifespan, death at 10 is worse than death at 90.

However, even if it puts the badness of death in the right perspective— neither overestimating nor underestimating its badness—McMahan's theory does not help us to answer the question about whether pursuing indefinite life extension would be worthwhile. So we need to continue our inquiry.

Epistemic Disagreement About Plausible Counterfactuals

Consider this sentence: had Epicurus never died, he would still be alive. This is, of course, a tautology. We can easily conceive of an alternate universe in which humans live for thousands of years, and in which Epicurus is still alive. However, when Epicurus died (in our current universe) in 240 BC, cryonics and life extension were merely conceivable, but not possible or probable.[3] Epicurus did not have a real possibility to live for more than 80 years, just as he did not have a real possibility to have wings. Both facts are bad if we assume that Epicurus would have enjoyed living for thousands of years and flying on wings, but they are not bad in the same way that the death of Epicurus at the age of ten would have been bad.

But how should we assess the badness of someone dying today, or ten years from now, as opposed to living for thousands of years? After all, if one considers indefinite life extension possible or even probable, the counterfactual life against which they assess the badness of death suddenly becomes extremely long and valuable. In this view, when a person of any age dies, the counterfactual life of which they are robbed measures not in years or decades, but in centuries and millennia. But as we have seen, counterfactuals need to be plausible if we are to take people's deprivation of them seriously, so these greatly increased stakes only matter insofar as they are plausible. Thus, we must look more closely at the arguments supporting the plausibility of life-extension technology.

All but the most pessimistic of critics would agree that our odds of someday attaining millennial lifespans are higher today than they were in 240 BC. A growing number of people today consider indefinite life extension as not merely conceivable, but realistically possible and, in some cases, even quite probable.[4] To date, however, most of the technologies required for indefinite life extension do not really exist in the same way that, say, in vitro fertilization (IVF) or quantum electrodynamics can be said to exist. Whereas IVF has produced hundreds of thousands of children, and quantum electrodynamics has given us transistors (which made computers resource-efficient and hence affordable), indefinite life-extension technology has so far only yielded a few hundred frozen corpses, some interesting proofs of concept, and a lot of hopeful extrapolation. But given the fact that a growing number of experts reckon life-extension technology will become available at some point in the not-too-distant future, it is not irrational to consider this outcome as at least being more probable today than it was in the past, even if it is not considered very probable in absolute terms. There are technologies that are not available yet, but that we are reasonably confident will be available in the next few years, and such confidence about their future availability affects their epistemic status. For example, it would be absurd to say that, since human clones do not exist yet, they must be as improbable as humans with wings. We are reasonably confident that—barring ethical or legal constraints—human clones are way more likely to be born within the next decade than, say, human-albatross chimaeras. However, some people think that indefinite life extension is as likely to happen as human cloning, whereas some other ones consider it as unlikely as winged humans.

The disagreement about indefinite life extension's expected odds of success is likely to produce disagreement about the badness of death at an old age, seeing as the counterfactual life that the cryonicist has in mind is significantly longer than the counterfactual life in the mind of the cryo-sceptic. From the cryonicist perspective, if someone dies at the age of 5, they lose about 80 years of probable life and some thousands (or more) years of possible life. The person who dies in their 80s has not lost many years of probable life, but they have also lost some thousands (or more) years of possible life. To the cryo-sceptic, instead, the five-year-old has lost many probable decades, whereas the 80-year-old has used almost the whole potential, and has lost almost nothing, given the very low chances that indefinite life extension will become available.

However, if deprivation of a future life is bad, then there is no reason why such deprivation should ever cease to be bad, even if the degree to which it is bad decreases. So even from the perspective of the cryo-sceptic, it is difficult to argue that death is not bad when it occurs at an old age. The fact that some people die at a very young age makes their death *worse* than the death of people who die at a very old age. But the fact that the people in the first group had it worse than people in the second one does not mean that the people in the second group did not suffer any harm. If life is always better than death, as Nagel suggests, then there is no limit to the number of years that it would be good to live, and there is no reason to think that the death of even decrepit people is not bad.

Death as Deprivation of Negative Counterfactuals

The problem with Nagel's conclusion that death is always bad because life is always good, however, is that the premise that life is *always* better than death, and that death *always* deprives us of something net positive, is at odds with the fact that sometimes people choose to die because their life is unbearable to them. Indeed, one's life might be unbearable and not worth living even if one does not choose to die (if one does not find the courage to commit suicide, say) and even if they cannot choose to die (such as with a neonate born with a serious and unbearably painful disease). Nagel builds his argument on the premise that the counterfactual life one loses by dying is a life one would want to live, even if it involved suffering. But people who choose euthanasia, for instance, choose death over a life of suffering. To them, death is better than life because the counterfactual life they would live (had they not opted for euthanasia), would be a life they do not want to live.

Not only can the badness of a counterfactual life make suicide or euthanasia rational; in some extreme cases, it can even make homicide an act of compassion. For instance, if I am involved in a car accident and my car catches fire while I am trapped in it, and I cannot be rescued, someone shooting me and sparing me the agony of a slow death would not be harming me; on the contrary, they would actually be doing me a big favour. Given an extremely bad counterfactual, death is not bad. So for death to be bad qua deprivation of life, we must assume that the life in question is at least a decent one. The cryonicist would generally agree that merely adding years to one's life is not valuable unless that life is at least worth living. This is why rejuvenation is a key step in the life-extension process and also in the perspective of being cryopreserved. Living for centuries without extensive rejuvenation (to the extent that it would even be physically possible) seems like a truly gruesome fate, as one's body and mind would slowly degrade with no guarantee of a merciful death around the corner.

The issue of negative counterfactuals, and the goodness of death against some extremely negative counterfactuals, opens up a new set of questions. At some point in the future, an agreement will likely be reached with respect to the possibility of achieving immortality; however, this would not solve the disagreement about the desirability of immortality. As the conservative motto goes, "just because something is technically feasible does not imply that it is also morally permissible". Two people could agree that death is bad, and that indefinite life extension is a real possibility, and yet disagree about the desirability of immortality.

Immortal people would have the obvious advantage of never being deprived of the goodness of life, and they would have many more chances to become what they want to be, to achieve their goals, to cultivate their relationships. But what if they found out that life necessarily becomes less good, and more burdensome, as time advances? What if there is something inherently boring, or tiring, or sad, or painful about existing for millennia and beyond? If so, we would have a different kind of reason to argue that cryonics and indefinite life extension are not a good investment. In the following section, we will explore this set of arguments.

DEATH AS FRUSTRATION OF DESIRES

As we have just seen, the problem with the deprivation account of death as formulated by Nagel is that it assumes that death is always bad because it always deprives of some net value (living). A different view based on a more modest formulation would be the following: *Death is bad if it deprives someone of a life they would have preferred to keep living.*

Once we introduce preferences into the picture, however, we have to abandon the assumption that death (qua deprivation of life) is intrinsically bad, and we need to rethink the badness of death as frustration of our preferences or desires to keep on living. The advantage of this preference-based approach is twofold: (1) It reconciles the intuition that death is normally bad with the intuition that it is reasonable for someone to choose death under certain circumstances; (2) it does not need to rely on some overly complex comparison of postmortem nonexistence and counterfactual existences in order to prove the badness of death. However, there are some difficulties also with this approach, as we shall now see.

According to philosopher Bernard Williams, there are at least two orders of desires: the conditional and the categorical ones (Williams, 1973). Williams argues that death is bad because it frustrates the categorical desires an individual might have. Before we delve deeper into this matter, let us first clarify the difference between these two categories.

Conditional desires are contingent upon the fact that one is alive. The desire to have food and water, for instance, is due the contingency of being a biological organism that requires food and water to survive. If the brain of such a being were uploaded onto a digital substrate, such as a computer or a robot, those specific conditional desires would disappear (and new ones, such as the desire to be connected to electric power, would perhaps arise). We do not normally consider it a pity that a recently deceased person will no longer be able to keep breathing air and drinking water. According to Williams, conditional desires do not propel our existence; they do not provide us with strong enough motivation to keep living. In other words, if all of one's desires were merely conditional, they would lack a motivation to continue to exist.

Categorical desires, meanwhile, are the ones that would (likely) be present regardless of one's substrate—that is, desires that are not merely contingent upon the factors that keep one alive. It is these desires that make us fear death more than anything. While it may seem like only grandiose desires would fit this criteria, this is not so; while Napoleon surely had a categorical desire to win each and every battle in his career as a military commander, his more modest desire to be loved and recognized by those closest to him was no less categorical. A less ambitious person may have the categorical desire to see the Great Barrier Reef, try all the ice cream flavours in the world, or learn to reverse-parallel park. Regardless of the specific contents, such desires motivate us—to different degrees—to keep living.

Bernard Williams famously discussed immortality and longevity in an essay published in 1973 entitled "The Makropulos Case: Reflections on the Tedium of Immortality". Williams introduced a fictional character, Elina Makropulos, who gives the title to the essay.[5] One day, Elina's father gives her an elixir that extends her lifespan by 300 years and that can be taken several times. She takes the elixir once, lives up to 342 years, and eventually decides not to take a second dose because, in Willliams' words, "Her unending life has come to a state of boredom, indifference and coldness. Everything is joyless: 'in the end it is the same', she says, 'singing and silence'."

Just like people who would use rejuvenating medicine in order to extend their life well beyond current boundaries and perhaps even achieve immortality, Elina Makropulos does not become decrepit, but stays biologically young for about 300 years. So in imagining her life, we should not picture a sort of Methuselah, unable to perform even the most basic of activities; otherwise her choice to refuse to take the elixir would be quite understandable. Moreover, when she decides to die at 342 years of age, she has not been living an extremely long life compared to what cryonicists and indefinite life-extension enthusiasts hope to achieve, so Williams' arguments would also apply to cryonics and rejuvenation techniques that fall short of the best-case scenarios envisioned in the context of cryonics and life extension.

Williams identifies three undesirable features in Elina Makropulos' life, all of which are meant to show why a very long or immortal life would be undesirable. We will start with lack of categorical desires in the remainder of this chapter paragraphs, before moving on to boredom and unrecognizability in the next chapter.

According to Williams, an indefinitely long life would be undesirable because all categorical desires would eventually be fulfilled, after which one would only remain with uninspiring conditional desires. Such a life would ultimately become unbearable due to the existential boredom, emptiness, and apathy that would come to pervade it.

Now, there are some problems with Williams' view. The first problem is that it is not obvious or uncontroversial that categorical desires are the *only* thing that propels us towards the future life. We can imagine a life wherein categorical desires are entirely absent, and yet the subject is happy and would rationally want their life to continue. Imagine, for instance, that a patient suffering from near-total paralysis decides to spend most of their life under the effect of some intensely pleasurable and side effect-free recreational drug. We would think that such a person has no categorical desires, but, nevertheless, given the well-being experienced thanks to the drugs, her life feels pleasant and worth living. It seems that, to Williams, such a scenario would be inconceivable because one could only want to keep living in order to fulfil their categorical desires; yet it is not difficult to suppose that at least some people could find such a life worth living.

Another problem with Williams' argument is that its implications are at odds with commonly shared views on the badness of killing people or animals who temporarily or permanently lack such desires, either because they suffer from depression or other mental disabilities or because they are members of species whose individuals do not have the mental capacity to formulate categorical desires. If categorical desires were all that mattered, we would have to consider it morally permissible to murder a considerable number of humans and even more nonhuman animals just for fun. People might have different views about the moral status of animals, and many people consider it is morally permissible to eat them in order to meet some nutritional needs or gustatory desires. Others might find it permissible to kill people with no categorical desires if they ask to be killed. But killing animals or some groups of people for no reason at all or just for fun is not morally justifiable within any widely recognized ethical frameworks (those who do find it permissible are generally considered to suffer from mental disorders like sadistic personality disorder or psychopathy).

Jeff McMahan (1988) points out that even though the badness of death cannot be entirely explained by the fact that it frustrates the victim's desires, such frustration is an important part of the reason why death is bad when it interrupts a life worth living; as he writes, "death frustrates the victim's desires, retroactively condemns to futility her efforts to fulfil them, and generally renders many of her strivings vain and pointless". Moreover, as we saw before, the badness of someone's death is measured against the quality and the quantity of goods they would have experienced if they had survived, and the way that one measures such loss depends in part on how much the subject wanted to have such goods or fulfil such desires. In this perspective, the death of a person who only has the weak desire to, say, try different ice cream flavours should be perceived as less bad than the death of someone with more complex and numerous desires (such as the strong desire to write a book, create a family, find a cure for cancer, etc.)

A third problem is that it is not obvious whether categorical desires would necessarily be depleted over a very long or immortal life. We can imagine an indefinitely long life during which categorical desires never disappear, but merely change object, intensity, or both. We can, for instance, imagine someone having the categorical desire to help their family throughout their whole life. Although the objects of that desire will inevitably evolve as various family members change, die, and are born, one would probably never lose the desire to be there for their family.

Overall, given that barely anyone has ever lived for much longer than a century, it is hard to predict whether Williams' assumption is correct. We cannot predict whether categorical desires would be depleted over a very long life.

A possible alternative may be to move away from the assumption that the only life worth living is the one sustained by categorical desires, and towards considering pleasure or even happiness as not necessarily associated to desires and to the satisfaction of those desires. Contemporary philosophers John Martin Fischer and Benjamin Mitchell-Yellin (2014) have argued that Williams appears to view life as a sort of library filled with a large yet finite number of books (a metaphor for categorical desires). Given an infinitely long life, one would eventually read all of the books in the library, and thereafter run out of projects. However, Fischer (1994) has also argued that some kinds of pleasures are *repeatable*: "such activities (and others) might well reliably (and repeatedly) generate experiences that are sufficiently compelling to render an immortal life attractive on balance". The kinds of activities that Fischer has in mind involve more basic pleasures, such as sex, food, listening to music, exercising, and so on. Unlike activities that stop providing pleasure once the set goal is reached, such as learning new skills or perfecting old ones, repeatable pleasures sustain their pleasurable effect over time. According to Fischer, then, an eternal life in which both repeatable and self-exhausting pleasures coexist would not become unbearable, because one would always experience the satisfaction of some desire—be it conditional or categorical.

In the end, it seems that death as frustration of desires by itself does not provide us with a convincing answer to our inquiry on whether death is good or bad. On the one hand, death is bad when it prevents someone from fulfilling their goals and desires to enjoy the goodness of life. On the other hand, the possibility that, given enough time, one would necessarily run out of those desires that propel us into the future casts a gloomy light on the perspective of immortality. In other words, even if death is a bad thing, it might be that immortality is even worse, as William's view implies. In the next chapter, we will take a closer look at immortality in order to understand if Williams' concerns and similar ones are justifiable, and whether there are arguments that can convince us that an immortal life would be so undesirable that pursuing indefinite life extension would be futile or even undesirable.

Notes

1. Given the limited scope of this book, I will leave aside non-Western cultures in which the influence of traditions like Hinduism and Buddhism have contributed to the development of a different approach to life and death.
2. Many languages have an equivalent of the sentence "if my grandmother had wheels, she would be a cart", a joke used to reply to someone fantasizing about hypotheticals in which the premises implies a scenario that is incompatible with the existence of the subject who is speaking ("If I were born in Ancient Greece" or "If my mother were Cleopatra").
3. I will consider *conceivability* the mere metaphysical possibility of something (I can conceive of a green horse, but I cannot conceive of a horse that is also a cabbage). I will consider possible something that is not merely conceivable, but also has some chances to happen (it is possible that a horse lives up to 50 years, although it has never happened). I will consider probable something that has quite high chances of happening (it is probable that a horse lives at least 20 years).
4. For instance, Google co-founders Larry Page and Sergey Brin are among those who think indefinite life extension will probably be available within the next 20 years. See, for example, Friend (2017).
5. This character was originally developed in a play by Karel Čapek, and later on adapted into an opera by Janacek.

References

Benatar, D. (2008). *Better never to have been: The harm of coming into existence*. Oxford University Press. Retrieved from https://market.android.com/details?id=book-paoVDAAAQBAJ

Fischer, J. M. (1994). Why immortality is not so bad. *International Journal of Philosophical Studies*, 2(2), 257–270. https://doi.org/10.1080/09672559408570794

Fischer, J. M., & Mitchell-Yellin, B. (2014). Immortality and boredom. *The Journal of Ethics*, 18(4), 353–372. https://doi.org/10.1007/s10892-014-9172-3

Friend, T. (2017, March 27). Silicon Valley's quest to live forever. *The New Yorker*. Retrieved from https://www.newyorker.com/magazine/2017/04/03/silicon-valleys-quest-to-live-forever

Kahneman, D. (2011). *Thinking, fast and slow*. Macmillan. Retrieved from http://www.math.chalmers.se/~ulfp/Review/fastslow.pdf

McMahan, J. (1988). Death and the value of life. *Ethics*, 99(1), 32–61. Retrieved from https://www.ncbi.nlm.nih.gov/pubmed/11653818

Nagel, T. (1970). Death. *Noûs*, 4(1), 73–80. https://doi.org/10.2307/2214297

Williams, B. (1973). The Makropulos case: Reflections on the tedium of immortality. In B. Williams (Ed.), *Problems of the self* (pp. 82–100). Cambridge: Cambridge University Press. https://doi.org/10.1017/CBO9780511621253.008

CHAPTER 4

The Immortality Conundrum

Abstract Hypothetical future treatments aimed at "rejuvenating" the body, thereby keeping it young and healthy for an indefinite amount of time, could offer a form of biological immortality. This chapter explores whether such immortality would come with downsides that would eventually make it an immoral or an undesirable goal. In order to assess the desirability of an immortal life, a few key questions must be addressed: What would an immortal (or indefinitely long) life look like? What kinds of benefits and downsides would it provide? Some authors have suggested that an indefinitely long life would be unbearably boring, and thus undesirable. It has also been suggested that an individual living an immortal life would not be able to recognize it as belonging to them in particular, or even as a human life at all. This chapter addresses these and other questions about the nature and desirability of an indefinitely long life.

Keywords Immortality • Indefinite life extension • Amortality
• Existential boredom • Existential tiredness

DIFFERENT TYPES OF IMMORTALITY

When people talk about immortality, they do not always refer to the same thing. A Christian or a Roman Catholic who believes in eternal life after death would normally associate the concept of immortality to that of an immortal soul in a metaphysical dimension. According to most religions, humans become immortal not by staying alive, but by dying. It is believed that an omnipotent and omniscient god assigns a place (in heaven, or if one is unlucky, in hell) to the soul, after which there is no way to exit eternity. We can consider this a "coerced" form of immortality, since it is not actively chosen by the individual, and is inescapable.

Another form of immortality, commonly portrayed in the context of science fiction or epic narratives, is that attributed to individuals who are invulnerable but can still choose to die if they so wish. This "chosen immortality", sometimes called *amortality*, is probably the one most of us would find desirable. One could go about one's life without having to fear death, yet knowing that one could exit life at any moment if one wanted to. Although some think amortality could be achieved through future technologies, it appears quite difficult to imagine a system that is simultaneously completely invulnerable to external threats yet able to be terminated at one's own request.

Some have suggested that amortality could be achieved by means of "uploading" a complete copy of the information in one's brain on to a digital substrate, allowing one's mind to be simulated in a more robust and energy-efficient medium (Cerullo, 2015). Roughly, the idea behind brain uploading is that the information stored in the brain can be copied and uploaded on a non-carbon-based substrate, such as, for instance, a computer or a robot or a hybrid between a biological and a digital entity. In theory, once a brain has been uploaded, its content could be backed up an unlimited number of times—making it very difficult to destroy every last copy of, and thereby "kill", that person. In an ideal scenario, the uploaded brains would always be able to terminate themselves (and eliminate all possible copies), just like it is always possible to wipe a laptop's storage and all of its backups.

Leaving aside potential problems with viruses or other technical complications, it appears that radically modifying core aspects of human nature (like our biological substrate) would introduce a number of possibly grave dilemmas. Not only would we need to address empirical concerns about the actual possibility of transferring information, but we would also need to address metaphysical questions about consciousness on a non-biological substrate, as well as about the nature of creatures that were born as humans and had turned into machines or hybrids (Minerva & Rorheim, 2017).

Both coerced and chosen immortality are quite different from the kind of immortality that is even remotely conceivable with our current scientific means. Cryonicists, for example, want to achieve indefinite life extension in the physical world, not in a metaphysical one, and they want to have the possibility of opting out of life if they so wish. We do not yet know whether future therapies aimed at extending human lifespan will succeed. However, we can postulate that, if rejuvenating therapies and technologies at some point become so advanced as to render it possible to live in a young and healthy body for an indefinite time, then immortality will become at least *virtually* possible. Of course, *in practice* accidents and incidents could still destroy a body, but one could *in principle* live forever by undergoing rejuvenating treatments a virtually infinite number of times. Moreover, we can assume that medical research will keep advancing, so that more and more conditions we now consider fatal will be easily cured in the future, thereby reducing the overall risk of dying from illness, trauma, or accidents. Such "virtual" immortality, commonly known as *indefinite life extension*, is what I will be referring to when we speak of "immortality" in this chapter (unless otherwise noted).

If one assumes that both cryonics revival and rejuvenation technology will someday be available more or less simultaneously, it is reasonable to also assume that revived cryonicists would choose to undergo rejuvenation at a sufficient frequency to remain biologically young and achieve virtual immortality in a physically young body. There are two main reasons behind this assumption. First, most people would probably want to stay young for medical reasons: ageing involves a number of biological processes that weaken the body over time, making older bodies more susceptible to injury and disease (Hayflick, 2000). So from a medical perspective, it would make more sense to use rejuvenation as a preventive therapy for avoiding age-related damage altogether, rather than having to cure individual diseases caused by ageing (Bostrom, 2005). Second, on a psychosocial level, biological youth seems to be the already preferred stage for many people. As they age, adults often try to "stay young" or "look youthful", and frequently compliment each other on looking younger than their actual age. The abundance of anti-ageing remedies available on the commercial market (not to mention the near-total lack of an equivalent market for pro-ageing products) suggest that people would choose to stay young through rejuvenation if they had the chance, and we should expect this preference for youthfulness to remain unchanged also in the future.

What matters when discussing the ethics of immortality is to understand whether what we consider "virtual" immortality would be desirable or not, and whether it would come with ethical downsides that would eventually make it an immoral goal. If the latter turns out to be likely, we would have good reasons to argue that the cryonics enterprise should be stopped. But if the most robust objections to immortality and extreme longevity do not stand our ethical scrutiny, then cryonics should no longer be looked at with suspicion, and even be supported and pursued.

In the next few paragraphs, I will examine the most common arguments against indefinite life extension and immortality, and explore possible counterarguments to each one.

What Would an Indefinitely Long Life Look Like?

In the previous chapter, we considered how death would cause nonexistence, deprive us of a future life, and frustrate categorical desires and preferences. We described what death is and what death "does" to living beings and how, assuming a decent quality of life, most people would prefer to continue to live rather than die. However, to say that one would prefer to live rather than die does not entail that one would want to live indefinitely. In other words, the badness of death does not imply the goodness of immortality. Immortality might be as bad or even worse than death. In this section, we will consider reasons why immortality might be considered undesirable, or why, conversely, it might be a goal worth pursuing, both from a moral and from a self-interested perspective. In order to assess the desirability of an immortal life, we need to answer a few key questions: what would an immortal (or indefinitely long) life look like? What kind of benefits and downsides would it provide? And would it be possible to conceptualize an indefinitely long life as the story of one individual, or would it always be a series of stories of individuals—closely related to each other, but not psychologically connected to one another in the same way as all the pieces of my story are connected to me as an individual?

Freedom from Regrets

Let us first consider what is probably the best aspect of immortality. If immortality were an option, we would never suffer the bad aspects of death: we would not pass to a state of nonexistence, where our identity dissolves and we cease to be the subject of any experience. In more practical terms, an immortal person would not have to worry about running out of time to do the things they want to do. Of course, we are talking about virtual immortality, so one would still have some anxieties related to the risk of dying. Still, we can imagine how an indefinitely long prospective life would come with less death-related anxiety than a life that lasts for a century at most. Even if accidents could still kill, a virtually immortal person would have virtually infinite time to reach their goals, fulfil their desires, develop relationships, and enjoy the good things that life can offer. Just like most people in their 20s—given the remoteness and the unlikelihood of death at that age—rarely worry or think as much about death as people in their 90s do, we can imagine that in a world where people lived indefinitely long lives, death would always appear to be a remote possibility, rather than an impending one.

Under these circumstances, we can imagine that a person would be freed from the anxiety of watching time passing by, empowered by the prospect of having much higher chances of achieving their goals and fewer regrets. As Todd May (2014) pointed out, an immortal life would be free from regret, as we would always have the time to try once more to achieve our goals, to be one thing instead of another one, and hence to avoid the sadness of having missed out on some opportunities. We would live our lives without feeling constantly torn by the irreversibility of most of our choices. We would not be regretful for what could have been, but never was. The necessity to choose one path at a crossroad is one of the hallmarks of our existence, from choosing to get up in the morning or lie in bed all day, to the choice of devoting time to one project or another, to that of being with someone or not.

It is important to point out that regrets at an existential level are different from day-to-day regrets, as the latter tend to have a small impact on our life, and we can easily make a different choice the next time. I might regret my decision to order pizza rather than pasta at a restaurant, but neither the decision nor the subsequent regret will make a significant difference in my life overall.

Existential regrets, on the other hand, result not only from a strong feeling that we have chosen the wrong path, but also by a realization that we no longer have the time to begin anew and choose something different.

As we grow older, the options available to us narrow down at a speed that many of us find uncomfortable, if not downright scary. Let us consider our professional life: before we choose what kind of career to pursue, we can imagine ourselves taking many different paths. One can imagine oneself as a lawyer, doctor, actor, pilot, piano player, pharmacist, geneticist, and so on. In reality, the range of options available to us is not as wide as we might think, and it is constrained by the fact that we have limited time and each of us has only a few natural assets. As we try to minimize the time and effort required to learn new skills in order to apply for jobs, we choose a career path according to our interests, natural assets, and other environmental factors. So, for instance, a person with good mnemonic skills would perhaps choose to pursue a career in the legal profession, as remembering thousands of pieces of information is a big part of law studies, whereas someone with a wide vocal range and perfect pitch would be inclined to choose a career in vocal performance, and so on. After we have chosen a career path and worked in a certain sector for a number of years, it becomes difficult to move to something completely different, and we find ourselves "stuck" in a certain profession. We can, of course, move up and down the corporate ladder within the company we work for, or move to a similar role in a different company; but becoming, say, a chef after having been a pianist is not very easy, although surely not impossible. There are people who manage to successfully change careers twice or even thrice during their life, but even two or three careers paths are a limitation when compared to the countless options we have before we start making professional choices, and that we would have if our lives were indefinitely long. It thus seems reasonable to argue that immortality, by keeping open the possibility of trying all of the options that we had originally discarded, would free us from this kind of existential regret.

However, it is not obvious that immortality would soothe a different kind of existential regret—one that does not originate from having picked the wrong profession, or partner, or resident city, but rather from always having to choose a path. Even if we are happy with the outcome of our choice, we are saddened by the thought of those counterfactual lives we would have had, if only we had made other choices—not necessarily better ones, just different ones.

Infinite time does not entail infinite options. Immortality would only offer the opportunity to try more paths than we currently do, but it would not be the same as being able to branch in different copies of ourselves and live different lives following different choices, nor would it be the same as rewinding time and making a different choice from the start.

To start with, one cannot "undo" some of the choices they have made in the past. For instance, once one has had sex for the first time, one can never be a virgin again, no matter how long one lives. In a broader sense, there are choices, experiences, and things that happen to us that shape the way we see the world, interpret information, and do the things we do.

For instance, there are at least some professional choices that shape our approach to the world in a way that would be perhaps impossible to "undo". As a moral philosopher, for instance, one gets into the habit of considering pros and cons of arguments, systematizing different views within certain moral paradigms, and formulating thought experiments to test the plausibility of certain arguments. Once one develops this (or any other) filter to process the world, it is difficult or perhaps impossible to remove such filter and approach the world in a different way. So even if a philosopher were to decide to become a social worker one day, they might not be able to remove their "philosopher filter", although they may be able to add a social work filter on top. Of course, this is not to say that there is anything intrinsically undesirable about having multiple filters through which to process the world[1]; it is merely a reminder of the fact that some options could be available just once even over an indefinitely long life.

There is another form of regret that cannot be prevented by adding more time to our lives, namely the regret of not being given an opportunity. Whoever has experienced the pangs of unrequited love knows very well that if someone is not in love with another person, it is very difficult to make them change their mind (or their heart). Of course, having more time would increase the chances that a certain person falls in love with another certain person.[2] But since it is neither certain nor likely that unrequited love for someone would, given enough time, eventually be reciprocated, it may be that immortality will not save one from the regret for not being given the chance to be with someone they love. Falling in love is not a direct result of the time and energy we invest in the project of winning someone's love (although it may play a part, of course). Perhaps in a billion years, one could learn to play the piano like Mozart. Perhaps after studying hard for a million years, one could find a cure for cancer. But there are things that cannot be achieved through hard work, that are out of our control, because they are the outcome of a random series of coincidences and other unpredictable factors.

In sum, it seems that immortality would eliminate some but not all forms of regret from our existence. However, even if some options were precluded because of other choices made in the past, an indefinitely long life would still leave far more options open than a short (i.e. normal span) life. Within the options available to each of us, we could explore more than a couple of career paths, try activities that require a lot of time to master, and have more time to enjoy our interests, hobbies, relationships, and long-term projects. We would still have some regrets, but we would have less regrets than we can currently expect to have on our deathbed. So it would seem that immortality qua freedom from some types of regret would be good enough to decide that an immortal life would be better than a mortal one.

However, freedom from regret could come at a high price. In the rest of this chapter, we will tackle other challenges that might arise for an individual living an indefinitely long life.

Personal Identity

As we saw in Chap. 3, philosopher Bernard Williams argued that what makes life worth living is wanting to fulfil one's categorical desires—those goals, projects, and plans that propel people to the future. According to Williams, such desires would become depleted over a very long life, and one would hence lose the interest and motivation to keep living. But he also considered the possibility that such desires could change in dramatic ways over a very long lifetime. If this happened, he argued, then one could no longer be considered the same person that they were when they had radically different categorical desires. But if the person living through centuries is no longer "me", then why should I take an interest in her survival? I would have no reason to say "*I* lived for centuries, *I* survived hundreds of challenges". In Williams' own words:

> The state in which I survive should be one that, to me looking forward, will be adequately related, in the life it presents, to those aims I now have in wanting to survive at all ... [S]ince I am propelled forward into longer life by categorical desires, what is promised must holdout some hopes for those desires ... at least this seems demanded, that any image I have of those future desires should make it comprehensible to me how in terms of my character they could be my desires. (Williams, 1976, p. 91)

Let us assume, for the sake of argument, that Williams is right in claiming that categorical desires are an essential component of our personal identity. It seems that, given this account of personal identity, we do not really need to consider the case of an extraordinarily long-living individual in order to conclude that there has been a radical change in the identity of that person. If I think of myself at the age of five (F^5) and realize that I had radically different categorical desires from the ones I currently have, I should conclude that I am now a different person from F^5. Hence, I should conclude that she (I) did not survive the change of categorical desires that occurred over the past 30 years.

Indeed, if I compare F^5 and my current self (F^{now}), I can clearly see that they have radically different categorical desires. F^5 wanted to be a singer, to have bread and Nutella at every meal, and to develop magic powers. F^{now} wants to continue being a philosopher, to finish this book, and to live in a house full of pets. Yet, I would not draw the conclusion that it was not worthwhile for F^5 to continue to live for 30 years past the age of five. Similarly, the fact that I will likely change my categorical desires over the next 30 years does not worry me as much as the possibility that I might die over the next 30 years. I have the intuition that I would rather be someone with radically different desires than no one at all.

So if Williams were right in arguing that a radical change of categorical desires brings about a change in personal identity, it seems to me that this would be a problem for most people past the age of ten or so, given how quickly categorical desires appear to change between the age of five and ten. As Martin Fischer (2012) points out in discussing the same passage of Williams, we identify with, care about, and judge desirable futures in which our categorical desires change considerably. Perhaps, Fischer suggests, what matters is that such changes in categorical desires are the result of certain processes (such as ageing), rather than others (such as being manipulated). One can surely agree that having one's own categorical desires modified through manipulation would be bad (because someone is breaching our autonomy), but this does not at all imply that under normal circumstances, we would prefer death to an indefinitely long life throughout which categorical desires are (naturally) modified.

Even though I disagree with Williams' view that persistence of somehow similar categorical desires is the requirement to maintain personal identity over time, I agree that, in order to assess whether an indefinitely long life could be desirable, I need to assess whether an indefinitely long life could be desirable *for me*. So the question is, if I lived indefinitely, would I be capable of looking at my past history and feel a connection to those events, or would I be alienated from my very distant past in the same way I am alienated from events happening to characters of TV soap opera?

As humans, we need to organize the events of our life within the frame of a narrative that can make sense to us. Our brain organizes our experience in a way that can make sense to us, in such a way that it appears to be a story. We understand (and therefore value) our lives not because we perceive them as a series of random experiences, but because we can connect these experiences to a *self*. In a sense, we see our lives as novels, rather than as a collection of short stories from different authors and about different characters.

According to a number of philosophers, personal identity is a crucial element of persistence over time and it is maintained over time through psychological continuity: one continues to be the same person not by continuing to have the same categorical desires, but by holding certain psychological relations between the experiences they have at a given time and the memory of such experiences (Parfit, 1984). The connections between the different stages of the self make life valuable, and death bad. The capacity to remember what has been processed by our brains, together with the perception that all these events are related to the same individual—the protagonist of the novel, essentially—are what makes each of us interested in continuing to exist as *this individual*. The desires, the projects, even the personality traits of the protagonist might evolve and change during the unfolding of the story narrated in book, but we can see that there is a connection between the character at page 10, aged 5, and the later character at page 200, aged 25. Moreover, in a coherent story, we see how some new desires have developed from previous desires, and how some events have determined a change of a certain trait of the personality because of being interpreted in a certain way by the protagonist.

But perhaps this kind of psychological continuity is also too high a requirement even for people with an average lifespan. Perhaps one does not need to be in an intransitive relationship of overlapping memories (or desires) with a past self in order to persist over time.

As we have seen, according to Williams, persistence over time is achieved through persistence of similar categorical desires, a clear sign that the person has not changed preferences, personality traits, and so on. According to other philosophers, what guarantees persistence over time is remembering to have been a past version of ourselves (although this too could be a tough requirement even for a 30-something like myself, not to mention for a Methuselah). Sophie-Grace Chappell (Chappell, 2009) suggests that Williams' concern about a given character's long-term persistence can be addressed by assuming that what matters for persistence over time is what Derek Parfit called "connectedness", rather than a thicker connection between the present self and the past selves as portrayed by Williams, which is what Parfit called "personal identity". According to Parfit's notion of connectedness, in order to persist over time, one would not need to be in an intransitive relationship of overlapping memories with their past self (for instance, I do not need to have overlapping memories with my five-year-old past self). Rather, it is sufficient that one's past selves are in a transitive relation with each other, and that they have overlapping memories. The crucial point here is that in order for me to say that I am the same person I was at age five, I do not need to remember what I wanted, how I felt, or what happened to me around the age of five. Instead, it is enough for me to remember what I wanted, how I felt, and what happened to me at the age of, say, 20. That version of me could recall what it was like to be me at the age of ten, and that version of me could, in turn, recall what it was like to be me when I was five. Thus, in order to guarantee my persistence over time for 30 years, I can rely on having overlapping memories with a somewhat younger version of me, and so on.

Compared to Williams' view, Parfit's view of overlapping relationships among past selves seems to present a more realistic requirement for retaining persistence over time, and for justifying one's interest in continuing to live for several more years. Personal identity is not a requirement to persist over time, in Parfit's view; hence the interest in my future self can be justified also in cases where I know that my future self will have no direct memory of my current self, insofar as there will be "intermediate" connections.

I think Williams is right to be worried about the survival of the protagonist's personal identity throughout a story that goes on for thousands of pages (and, in the case of real people, thousands of years). However, I think this concern is motivated not by a fear that the character's categorical desires would change, but rather by a fear that the character's brain might fail to keep together the pieces of the story.

Cryonics and life extension are not about keeping the same *body* alive indefinitely, but about keeping a certain *person* alive indefinitely. If the content of my brain were emptied to leave room to new memories, experiences, and so on, I would no longer be myself in a meaningful sense. I am only myself insofar as I can remember my past self and connect it to my present and future selves. Without this connection through stages of the same self, immortality is an empty concept, reduced to the immortality of a body. For immortality to be worth it, one must at least have some personal continuity and a relatively strong connection to some more recent versions of oneself—otherwise, one would not even be able to recall one's categorical desires (or projects, or plans) for long enough to fulfil any of them.

We do not know how long a human brain might be able to maintain an unbroken connection between past and present. If at some point (say, after 10,000 years) every human brain started to lose the capacity to provide continuity or connectedness with the past, then we would know that we have reached the page limit of a coherent human biography. Until we reach that point, though, it is reasonable to argue that we have an interest in continuing to exist.

A Recognizably Human Life

It is difficult to conceptualize the life of a creature who is virtually immortal, yet does not age.[3] If such a person appeared before us right now, we would probably perceive them as so profoundly different from us that we might struggle to recognize them as human at all (we would perhaps think he or she is more like an Elf, the eternally young human-looking creatures in *The Lord of the Rings*). Following Fischer (2012), we can call this question—whether someone living an indefinitely long life would still be recognizable as "human"—the "recognizability issue".

A number of arguments have been put forth to suggest why an immortal life would not be genuinely human. For example, Bernard Williams (1973) argued that a person living an extremely long life would lack a coherent human character; Martin Heidegger (1927) thought that a life without a clear end would be "formless" and therefore not recognizably human; and Martha Nussbaum (1994) argued that our fundamental values would be deeply affected by extending our lives indefinitely.

One aspect of the recognizability issue that appears to be particularly interesting when considering life extension paired with rejuvenation is that of the separation between chronological and biological age. As we know, the immortality that cryonicists want to achieve involves extending a state that mimics biological youth/adulthood for an indefinite amount of time. If such a goal were achieved, ageing would become something radically different from what it is now, as there could be an unprecedented discrepancy between an individual's chronological and biological age: someone could appear to have a biological age of 20 years, while being chronologically 20,000 years old.

At the moment, we have a relatively clear idea of how a person's looks should correspond to their age and, in turn, what stage of life they have reached. For instance, we think of people in their 20s as looking very youthful, and devoting their lives to studying, entering the job market, and partying. We imagine people in their 30s and 40s as less youthful-looking, and more focused on consolidating their careers and starting a family. People in their 60s and 70s tend to have wrinkles and grey hair, have already achieved their professional goals, and enjoy greater economic and psychological stability than younger people. Of course, these are generalizations, and we all know someone who does not conform to the stereotypes of their age group. Nevertheless, statistical tendencies do exist (which is why a random sample of nightclub patrons will consist mostly of people in their 20s and 30s), and they do influence our social expectations.

So, if an indefinitely long life necessarily undermines this correspondence between biological and chronological age, does that mean the life necessarily loses its "human" character? And if so, would it be a bad thing? Human life as we know it progresses through different biological and psychological stages that are strictly related to each other—but would modifying or completely erasing such stages make it less human? On the one hand, we need to understand whether biological ageing is necessary in order to age psychologically and experience different stages of the human life cycle. On the other hand, we need to understand whether there is something valuable in experiencing all the different stages of life that we currently experience and in the particular way that we experience them.

Let us start with the question of whether, in order to age psychologically, it is necessary to also age physically and biologically. As noted by philosopher Samuel Scheffler (2013), we understand human lives as made of stages, and we attribute certain goals, activities, and behaviours to each stage. We evaluate accomplishments according to the stage of life one is at: we consider it an accomplishment for a 2-year-old to be able to speak, but not for a 30-year-old to do the same. We need to refer to different stages of life in order to decide what goals to pursue, what activities to engage in, and to accurately assess our own achievements. When I evaluate myself, as I think most people do, I estimate the percentage of my expected overall life that I invested in a certain project. I compare myself to other people my age, and I remind myself I am not just racing against other people, but mainly against the limited time of my life. As Todd May (2014) put it:

> For mortal beings, as we have seen, life is fraught. What happens is fragile. It might not have happened ... The events of our lives, both good and bad, have an urgency that they do not have for the Immortals. An Immortal does not worry about missing anything. There is time for it to be experienced ... For those of us who die, there is a singularity to the moments of our existence.

So not only do the different stages provide a direction in which to take steps and achieve goals, but it is the whole finitude of our life that gives our life a frame within which some things are valuable and some other things are not. If I want to climb Mount Everest and can only do so before, say, the age of 50, then I know that climbing Mount Everest is a priority during a certain stage of my life. Over a mortal life throughout which I age, I would know that goals requiring intense physical effort should be achieved by my late 20s, goals requiring intellectuals effort should be achieved when my intellectual capacities are at their peak (say, between the age of 30 and 60), and projects requiring minimal physical and intellectual effort should be saved for a time when I am old. But if I can climb a mountain at any chronological age, write a book at any chronological age, spend five years watching TV at any chronological age, then—assuming that I would like doing all of these things—I would have no criteria for prioritizing one project over the other.

In this sense, extending life indefinitely would move, or perhaps even dissolve, all of our current goalposts. It seems undeniable that increasing the human lifespan while reversing the ageing process would drastically modify our understanding of life as a series of distinct stages we usually go through. Perhaps there would still be stages, but because the correspondence between biological and chronological life would be deeply altered, the stages would be profoundly different from those that characterize "human" life as we know it. Admittedly, though, it is very difficult to understand why people tend to behave according to conventionally defined life stages. I think that at least three plausible hypotheses can be put forward to explain this tendency:

1. People behave in ways considered appropriate to their stage of life because of the physical and physiological characteristics that are associated with a certain age. For instance, people often stop clubbing in their 30s because their body gets older and struggles to keep up with dancing and partying for several hours in a row. So, if it were the case that people behave according to different life stages because of biological ageing, then it would seem that eternal youth might imply being stuck at a certain stage of life indefinitely (such as the stage of life at which people usually enjoy clubbing).
2. People behave in ways considered appropriate to their stage of life because of societal expectations about age-appropriate behaviours. Perhaps people stop clubbing because there is social pressure to stop clubbing at a certain age, and people who keep clubbing well beyond their 20s are frowned upon. So, if it were the case that people behave according to different life stages because of social pressure, it would seem that eternal youth would give people a way to avoid the pressure of having to move to the next stage of life. First, people around them would probably have no idea about what kind of stage of life is appropriate to a person who is, say, 2000 or 10,000 years old.

Second, as nobody would be able to distinguish between a 2000-year-old and a 20-year-old, no one would frown upon them if they did not behave according to their age.
3. People behave in ways considered appropriate to their stage of life because they change preferences after having experienced something a certain number of times. Perhaps people stop clubbing because clubbing becomes uninteresting and boring after one has clubbed for a certain number of years. So if it were the case that people behave according to different life stages because most things become boring after a while, and all people start trying them around the same age, then eternal youth would not keep people stuck at the same stage of life forever, as people would nonetheless get bored and move to the next stage. Perhaps there would be whole new stages of life we cannot imagine yet—a stage at which one buys a house on Mars, say.

Let us move on to the second question: we need to understand whether there is something valuable in experiencing all the different stages of life that we currently experience, and whether we would stop being recognizably human if we were to remove these stages.

In an internal report on ageing, the President's Council on Bioethics (2003) argued as follows:

[W]e must take into account the value inherent in the human life-cycle, in the process of aging, and in the knowledge we have of our mortality as we know it. We should recognize that age-retardation may irreparably distort these, and leave us living lives that, whatever else they might become, are in fundamental ways different from—and perhaps less rich than—what we have to this point understood to be truly human.

It is possible that members of the President's Council on Bioethics are wrong in assuming that there is something valuable in experiencing life as a cycle, with its succession of stages as we know them. Perhaps this is a fallacy similar to the one we discussed in the first chapter, when we saw how most people assume that the perfect age to become a parent is around the early 30s—which just so happens to be just before the time a woman's fertility decreases significantly. We saw how in vitro fertilization (IVF) and other fertility treatments contributed to change this assumption, and that nowadays fewer and fewer people think that there is any normative value in biological fertility. Similarly, there may be no particular reason to assume that there is something morally good in becoming biologically old(er). Perhaps a good life does not necessarily involve biological and psychological ageing, but can consist of a long, perhaps eternal, youth.

However, one could argue that, even if there were no value in maintaining a particular structure in the succession of stages of life, there might be something valuable about experiencing all the stages of human life. Perhaps there is something valuable in being old, and if immortal beings never reach that stage of life, then immortality would deprive us of a valuable experience. Perhaps we feel sad when a young person dies not only because they were deprived of a certain number of years, but also because we regret that they did not have the opportunity to experience all of the different stages of life.

It is not possible to say whether life extension would make people miss the opportunity of experiencing some stages of life. On the one hand, it is plausible that growing older without ageing biologically would dramatically affect the way we experience the passing of time. If a large part of ageing has to do with looking older and being perceived as older by the people around us, then by extending life through rejuvenation, we would affect the unfolding of our life cycle. If in order to experience, say, the calmness of old age one needs to actually *feel* that their body is becoming weaker, then it is reasonable to foresee that a person that never ages biologically, regardless of their chronological age, would never experience life as an old person and the calmness that comes with it.

But, on the other hand, it is also possible that the passing of time leaves a deeper psychological mark in our life, and that ageing brings about some radical psychological changes even if one's body does not age at all. If ageing has mostly to do with gaining experiences, learning from mistakes and successes, and developing psychological tools that allow us to navigate the world more safely—gaining wisdom, in other words—then the fact that our body looks and feels young should not have much of an impact on the process of ageing.

In sum, it is hard to predict whether indefinite life extension and rejuvenation will dramatically change the structure of life as a series of distinct stages humans usually go through. It is not even clear whether the concept of stages of life would simply just disappear once indefinite life extension became a reality. At this point, we have no idea whether ageing biologically is a necessary step for ageing psychologically. Moreover, we do not know whether ageing psychologically is valuable per se, or whether it is valuable only in the context of a finite life.

Within all this uncertainty, it nevertheless seems fairly certain that, from the perspective of human creatures like you and me, it would be difficult to recognize a life that does not go through these stages as a human life. However, a life worth living does not have to be necessarily human. We saw in the previous chapter how what matters in order to attribute a value to our own life, and to have an interest in continuing to live, is to be a *person*, not a human being. It seems to me that there is no reason to doubt that someone born human and subsequently made immortal would not cease to be a person.

Would Virtual Immortality Deprive People of Eternity in Heaven?

Up to this point, we have explored some implications of an indefinitely long life by considering death as the only plausible counterfactual. This is because, from a secular perspective, one ceases to exist as a person and subject of experience as soon as one dies. But most belief systems worldwide tend to include eschatological claims about the supposed persistence of subjective experience and personal identity after death, usually in the form of either an afterlife in a separate metaphysical dimension or some form of spiritual reincarnation in this one. Although I am personally sceptical about claims of this sort, they represent most people's ethical beliefs about death and form an integral part of most societies worldwide, so no ethical analysis of death would be complete without taking them into account. Hence, we will now explore some implications of cryonics and indefinite life extension in this context.

According to some religions, including the three most populous ones worldwide (Christianity, Islam, and Hinduism), one's soul continues to exist indefinitely after death.[4] The kind of immortality featured by these religions is what I have described as coerced immortality. I would argue that coerced immortality is not desirable because one could become bored and tired of life even if it was pretty good and varied, but especially because we cannot rule out the possibility that one's life could become extremely painful, and, in that case, to experience pain for eternity would surely be extremely undesirable. If one had to suffer an eternity of what Dante Alighieri described in the first part of his *Divine Comedy*, one would surely be better off dead (i.e. nonexistent). Of course, for people who hope to end up in a more heavenly scenario, where they experience eternal bliss and contentment, the prospect of immortality looks far more appealing.

Unlike people who would try to achieve eternal life through brain uploading, immortal souls would not have to worry about the risk that their brain might get hacked and tortured. And unlike people who would try to achieve eternal life through cryonics and rejuvenation, immortal souls would not have to fear the possibility that a major accident or an incurable disease could kill them.

It seems, then, that people who have good reasons to believe that an eternally blissful afterlife awaits them after death would have no reason to try to achieve immortality through medical and technological means. Indeed, they would have no particular reason to even extend their own lifespan beyond what has been assigned to them by God. If someone were sure that there is a God who will provide them with eternal bliss, it would be irrational for them to pursue not just immortality, but even moderate life extension. Life on Earth surely is not constantly blissful; on the contrary, it involves significant amounts of pain and suffering even for the luckiest of people, let alone those who live under particularly difficult circumstances. So if Heaven were a likely alternative to Earth, even the luckiest people would have good reasons to keep life as short as possible. Considering that suicide is considered impermissible by a number of religions and hence would not work as a shortcut to Heaven, it would make sense for religious people to keep the length of their life to a minimum. However, it does not seem that religious people are less keen than non-religious ones to use life-saving medications, from simple antibiotics to complex surgeries. Indeed, most people, including religious ones, are interested in extending their lifespan.

There are some possible explanations for this apparent lack of eagerness to pass on to the afterlife. One explanation could be that, even though religious people believe that the afterlife will be better than their mundane lives, they are nevertheless scared of death. If so, it is unlikely that such fear would decrease over time, as they get older and older thanks to rejuvenation. If anything, it is possible that, since death will be perceived as increasingly alien to humans (as people die at a way slower pace or do not die at all), the idea of dying will be perceived as even scarier, because death will be less of a common experience.

Another explanation is that they are scared that they will not be awarded eternal bliss, and they will end up in Hell. Even though most religions prescribe specific behaviours, there is always room for uncertainty about how God could judge one's overall conduct. From this perspective, people might try to live long enough in order to have the chance to prove to God that they deserve to be in Heaven. If in the current scenario this trial would normally last, at most, one century, it is possible that in a future where the average life is way longer, people will feel more anxious about trying to be good and secure themselves a place in Heaven. So they might feel compelled to keep trying for several centuries while conducting increasingly virtuous existences.

Religious people may also have some doubts about the existence of God and the afterlife, and thus want to make the most of their mundane life in case there is nothing in store for them after they die. This doubt is not likely to be alleviated with more time, so if one is motivated to use current life-saving treatments based on the doubt that there is nothing better waiting for them in the afterlife, they might end up using rejuvenating techniques for way longer than they had planned—perhaps even indefinitely.

It is also plausible that since religious people conceive life as a gift from God, they want to show their appreciation by making the most of that gift by trying to avoid death and extending their life. Again, if a person wants to show appreciation for the gift of life, it is not so obvious that they should stop showing such appreciation as they approach the natural lifespan of humans. One could argue that this "natural" limit reveals the measure of the ideal lifespan according to God. But what is considered the natural human lifespan in a Western country is quite different from what is considered the natural human lifespan in a developing one, and the natural lifespan of people living today is longer than the natural lifespan of people living even just a few centuries ago. As we modify the environment surrounding us by removing dangerous predators from inhabited spaces, or by developing aqueducts that bring drinkable water to everyone, or by developing glasses for people with short eyesight, or intensive care units for people suffering severe accidents, we change the "natural" lifespan of humans. In this sense, rejuvenating technologies will not represent a novelty, but a simple continuation of the human project aimed at making human life more comfortable and longer.

Finally, both religious and non-religious people live closely connected to others, and feel responsible towards people dependent on them and worry about the impact of their death on loved ones. So even if one were really confident of the goodness of the afterlife, they would still have a good reason to try to stick around as long as possible in order to support people close to them. If most people keep living indefinitely, the feeling of responsibility towards them will also be extended indefinitely, as will the feeling of wanting or having to live indefinitely.

Regardless of why religious people try to live longer despite believing that they could be better off by dying and by ascending to Heaven, it seems that the option of an indefinite long life could translate into a difficult decision for them to make. If and when rejuvenating technologies will be used as preventive medicine, many people will undergo rejuvenation in order to avoid, for instance, to get cancer. But as more and more people will choose rejuvenation, the average lifespan will become noticeably longer. Eventually, religious people would find themselves in the awkward situation of having to choose between immortality on Earth and the prospect of immortality in Heaven (or in Hell).

Even though this is not a good enough reason for not pursuing indefinite life extension, it is worth considering that, if it turned out that Heaven actually existed, people living an indefinitely long (and suboptimal) life on Earth would end up missing out on a much better alternative.

Boredom

The most common argument against immortality or extreme life extension is based on the idea that life, after a certain point, would necessarily become boring (cf. Fischer & Mitchell-Yellin, 2014). One of the first philosophers to introduce this objection was Bernard Williams who, as we saw in the previous chapter, argued that a long life would become necessarily boring because of the depletion of categorical desires that usually propel people into their future.

Other philosophers have highlighted the issue of boredom as an inescapable feature of an immortal life. Shelly Kagan has argued that after a certain number of years, be it centuries, millennia, or billions of years, one would necessarily grow bored with life. According to Kagan, over a very long life, one would eventually achieve most of their goals, or become disengaged with the issues that used to keep them interested. For instance, someone with an interest in mathematics would eventually become bored with mathematics, even if there were still mathematical problems they had not tackled yet. So, in Kagan's view, boredom does not seem to necessarily develop from exhausting categorical desires, as Williams argued, but rather from losing interest in everything. An immortal human being would necessarily become disengaged with the world (Kagan, 2012).

Similarly, Todd May (2014) suggested that, after a certain point in a very long life, one would become bored of learning, practising, and improving their skills. May gives the example of a musician: musicians usually need a lot of practice in order to become very skilled, so an immortal musician could practise for much longer and thus become better at playing than any mortal musician. But for how many centuries could someone keep playing music before they stopped feeling like they are improving, or stopped enjoying it? May says that even though we could not be sure that a musician with infinite time to perfect his or her skills would necessarily become bored of playing, there are good chances that this would happen at some point: "Even the deepest of passions is likely to fade with the passage of enough time. What was not eroded by decades will probably be eroded by centuries or millennia" (ibid., p. 62).

Before we delve deeper into the boredom issue, it is important to note that there is a key difference between boredom in an ordinary sense and boredom in an existential sense. Although they are often conflated in the literature, the term "boredom" can refer to two possibly related but not overlapping states: I can be bored in an ordinary sense by things that are repetitive (I am bored because the customer service I have called has forced me to listen for 40 minutes in a row to the same jingle), without being bored at an existential level (I feel genuine enthusiasm for new experiences in my life). Conversely, I can be bored at an existential level (I have already had new experiences so many times that experiencing something new itself has become monotonous to me) without being bored in an ordinary sense (I am constantly entertained by videos of cute animals).

We are only concerned with existential boredom, as that is the kind of boredom that, according to some philosophers, would make a long or immortal life undesirable. Keeping the two concepts completely separate is not possible, because a life that was almost constantly ordinarily boring would sooner or later turn into an existentially boring one. Similarly, if one felt deeply existentially bored for a very long time, they would also probably be bored in the ordinary sense.

Consider now the following hypothetical scenarios of individuals living indefinitely long lives:

Scenario 1: Alice has lived for a million years: she wakes up at 7 a.m. every day, except on weekends. She has worked at the same company for 800 years, doing more or less the same set of tasks. She has always spent time with the same friends from childhood, and she has had the same partner through her whole life. She goes on holiday twice a year, usually to Italy and California, and likes to spend her free time reading novels and singing in the choir.

Scenario 2: Albert has also lived for a million years. He has travelled around the world several times, and explored all of its corners. He has visited the same cities many times, and has seen them changing in radical ways. He has read millions of books, some of them several times. He has watched millions of movies, TV shows, and concerts. He has learnt to play the piano, and when he became very good at it, he learnt to play the cello, and when he became extremely good at it, he learnt to play the trumpet, then the drums, and then the violin. He can now play most musical instruments, and he has played every piece of classical music ever composed at least once. He has changed jobs hundreds of times; he has been a plumber, philosopher, pilot, chef, brain surgeon, gardener and many things more. He has raised his children, and grandchildren, and great-grandchildren, and so on. He got married hundreds of times, and he has been married to the same person for thousands of years before moving to the next partner.

Scenario 3: Adrian has also lived for a million years, and he has been connected to experience machines for a large part of his life. He has lived in a simulation where he has experienced life not just as himself in countless parallel universes, but also as a bat, an alien, a dinosaur, an ocean, and countless other things. He has also experienced the world as different people, including historical characters like Napoleon and Marie Curie, as well as fictional ones like Anna Karenina and Harry Potter.

The first scenario seems to be the most boring in the ordinary sense (i.e. repetitive and monotonous) because Alice would go on living the same life for an indefinitely long time. However, it would appear that, at least in the second and third scenario, objections to immortality based on "ordinary" boredom would be hardly justifiable. If monotony and repetitiveness are what cause ordinary boredom, then the second and especially the third scenario would appear to describe a desirable indefinitely long life. However, I think that there are two considerations that ought to be taken into account before getting to the conclusion that scenarios 2 and 3 are the least boring, and that they would be the most desirable.

First, it is plausible that, because of different psychological makeups, different individuals would perceive the same kind of life (repetitive as in scenario 1, and increasingly exciting in scenarios 2 and 3) in the same way.

Alice, for instance, might not feel bored in an ordinary sense. She might enjoy spending her life in her small town, within her small social group, doing the things she likes to do and the ones she has to do. Unless one's life were repetitive to an extreme level, for instance, if one were forced to watch the same 30 seconds of the same video while eating boiled potatoes for eternity, repetition and monotony would not necessarily cause someone to be bored either in an existential or in an ordinary sense. Meanwhile, people like Albert or Adrian might need to change their surroundings, their job, and their social context quite regularly. They might be the kind of people who need to travel very often, live in big cities, and try different jobs and partners throughout their life.

So it is possible that two people conducting very similar lives would have very different opinions about whether such lives are boring (in an ordinary sense) or not. Moreover, it is possible that even if Alice and Adrian agreed that Alice's life is boring, they would disagree on the assessment of the quality of Alice's life. Alice might think that although her life is boring, it is also peaceful, calm, and happy, and hence worth living. Adrian might think that Alice's life is so boring that, if he were in her shoes, he would prefer a different kind of life; and if that were not an option, he would prefer death over a boring life.

Second, even if the problem of ordinary boredom were solved, we would not be able to predict whether existential boredom would eventually arise, regardless of the differences in the scenarios we have considered.

It is difficult to predict whether all unnaturally long lives are bound to become existentially boring. In one sense, many of us begin to lose enthusiasm towards life long before they even reach a naturally old age. At the beginning of life, everything is new. There is the first time we pet a dog, the first time we make a friend, the first time we eat chocolate, hold hands with someone, swim in the sea, and so on. The goodness of novelty is not just about the content of each new experience, but also about experiencing novelty in itself. However, as we age, not only do the number of new experiences we can possibly have begin to narrow down, but we also get used to experiencing new things. There is a sense in which experiencing something new becomes obsolete, because we are familiar with the feeling of being surprised, of exploring something unknown, of being in a different situation. Yet even though life becomes a bit less "new" in this sense— and I think this is what existential boredom is—we still enjoy it for several decades. It is hard to predict whether, over a very long life, we would get so used to experiencing things that we would become simply incapable of experiencing any kind of emotion about anything—and whether, in such a state, we would prefer death over continuing to live.

Tiredness

It seems that some of the concerns about existential boredom would be better expressed as concerns about the "existential tiredness" that would arise over an indefinitely long life. For instance, Charles Hartshorne (1958) wrote:

> No animal endowed with much power of memory ought to live forever, or could want to, I should maintain; for the longer it lives, the more that just balance between novelty and repetition, which is the basis of zest and satisfaction, must be upset in favour of repetition, hence of monotony and boredom ... As Jefferson wrote to a friend: "I am tired of putting my clothes on every morning and taking them off every evening." (qtd by Lamont, 1965)

The Thomas Jefferson quote at the end of this passage is used by Hartshorne to support his claim that the longer one lives, the more repetitive, monotonous, and boring life becomes. However, Jefferson explicitly says is that he is tired, not that he is bored, and it seems to me that there is a relevant difference between being bored of life and being tired of living.[5] One could experience ordinary or existential tiredness without simultaneously experiencing ordinary or existential boredom (in the same way as I can be tired of walking without being bored of walking, and vice versa).

Just as different people might get bored more easily than other ones, some people might get tired more easily than others. Differences in how easily or quickly one grows tired might depend on different levels of resilience, or the intensity with which people live their life, or on the different challenges people meet over the course of their life.

Philosopher Peter Singer (1987) described life as an uncertain voyage. Whilst Singer in his essay focuses on the uncertainty about when life's voyage starts and ends, I think the metaphor also works perfectly to describe life in general. Life is a trip we all must take once we are born and as long as we choose to stay alive. Since both the path and the traveller are different, so, too, is each trip. The tiredness of carrying oneself through life is something that people could experience regardless of how beautiful and happy their life is. Philosopher Walter Kaufmann (1959) put it thus:

If one lives intensely, the time comes when sleep means bliss. If one loves intensely, the time comes when death seems bliss ... The life I want is a life I could not endure in eternity. It is a life of love and intensity, suffering and creation that makes life worthwhile and death welcome.

It may be that living intensely is a necessity for people who get easily bored, as only by living an intense life can they outrun the boredom that seems to always be at their heels. But there are also people who perceive their life as intense (and tiring) not because of the extraordinary adventures they choose to embark on, but because of their psychological makeup. Some people just happen to feel everything more intensely than others, and to them, the world is like a high-contrast photograph: bright colours are always extremely bright, black is always pitch black, and experiences make them either very happy or very sad. Life through high-contrast lenses has a higher emotional impact and is more tiring. There are also people with the opposite psychological makeup, who see life's colours as an uninspiring mix of pale hues and grey tones, as though they were always wearing dark sunglasses. And of course, there is a whole scale of intensity between these two extremes. I suspect that the more intense the world feels, the more tired of living one becomes, and vice versa.

And to some people, life is just very challenging. They have to push especially hard through their life because of illness, or poverty, or because tragic events have studded their path, making the trip more exhausting to them. Again, each trip is different, with obstacles of a different nature, and different travellers might be affected in different ways by the same type of obstacle.

Existential boredom is not the only possible threat to an indefinitely long life, because existential tiredness could arise even in a non-boring existence. However, to say that existential boredom and tiredness could arise over a very long life is not the same as saying that they would be a necessary feature of every indefinitely long life. And even if it turned out that many people living for a long time would eventually become existentially bored and tired, we should not assume that they would all choose death over life. A boring life might still be worth living, and a person who is tired of living might still want to keep going.

Of course, there probably is a limit to the degree of boredom and tiredness one can endure, and some people would probably choose to die, eventually. However, to say that existential tiredness and boredom might pervade the existence of some, or the majority, or even all of the people who choose to extend their lives does not imply that trying to extend human lifespan well beyond its current limit would be useless. We would not say that saving the life of a three-year-old is pointless merely because she might nevertheless want to die when she is 100. For the same reason, developing technologies that would allow people to live for hundreds or thousands of years would not be pointless simply because people might choose to die around their 9834th birthday. Insofar as we think our life is worth living, and dying would be harmful, the effort of extending a life is not pointless. Drawing a line at the current human life expectancy and confidently declaring that "This is the age at which dying is no longer a harm, but a blessing!" is as fallacious as drawing a line just before a woman's fertility starts to decrease, and declaring that to be the ideal time for her to have a child. We have adapted to these natural deadlines because we could not change them. We have internalized them and structured our lives and expectations around them. But when we are presented with an opportunity to push away our naturally imposed deadlines, we should not shy away, but rather ask ourselves whether it would be morally acceptable to take the opportunity, and whether the expected payoff would be worthwhile.

In the first four chapters of this book, I have considered various reasons for and against cryonics as a fundamental step in the quest for indefinite life extension. Although there is uncertainty about whether true immortality might be desirable, and about how long the ideal lifespan would be, I believe that adding time to our currently short lives would be desirable, morally permissible, and beneficial to many people—including those who are already very old. A trip of 80, 90, or even 100 years is not long enough; living takes a lot of practice, and death can wait.

Notes

1. In a sense, we already use multiple filters: I filter the world as a philosopher, as a woman, as a European, and so on.
2. *Love in the Time of Cholera* by Gabriel Garcia Marquez (1985) is a beautiful book about unrequited love that eventually turns in reciprocated love, after several decades of waiting.
3. Virtual immortality exists in nature. Such "biological immortality" is especially well known among members of the phylum Cnidaria, the ancient group of animals that includes the jellyfishes, corals, and comb jellies. These

relatively simple animals are able to regenerate all parts of their tissue, and a number of them—including the ubiquitous *Hydra*, which inhabit freshwaters worldwide, and the aptly named "immortal jellyfish" of the genus *Turritopsis*—do not show any signs of ageing over time.
4. It should, of course, be noted that views on the eternal life of the soul differ greatly between the Abrahamic religions and Hinduism.
5. By tiredness I mean a sense of psychological or physical exhaustion.

References

Bostrom, N. (2005). The fable of the dragon tyrant. *Journal of Medical Ethics*, *31*(5), 273–277. https://doi.org/10.1136/jme.2004.009035

Cerullo, M. A. (2015). Uploading and branching identity. *Minds and Machines*, *25*(1), 17–36. https://doi.org/10.1007/s11023-014-9352-8

Chappell, T. (2009). Infinity goes up on trial: Must immortality be meaningless? *European Journal for Philosophy of Science*, *17*(1), 30–44. https://doi.org/10.1111/j.1468-0378.2007.00281.x

Fischer, J. M. (2012). Immortality. In B. Bradley, F. Feldman, & J. Johansson (Eds.), *The Oxford handbook of philosophy of death*. Oxford: Oxford University Press. https://doi.org/10.1093/oxfordhb/9780195388923.013.0016

Fischer, J. M., & Mitchell-Yellin, B. (2014). Immortality and boredom. *The Journal of Ethics*, *18*(4), 353–372. https://doi.org/10.1007/s10892-014-9172-3

Hartshorne, C. (1958). Outlines of a philosophy of nature. *The Personalist*, *39*(4), 380–391. https://doi.org/10.1111/j.1468-0114.1958.tb03174.x

Hayflick, L. (2000). The future of ageing. *Nature*, *408*(6809), 267–269. https://doi.org/10.1038/35041709

Heidegger, M. (1927). *Sein und Zeit* [Being and time]. (J. Macquarrie & E. Robinson, Trans.). New York: Harper and Row.

Kagan, S. (2012). *Death*. Yale University Press. Retrieved from https://market.android.com/details?id=book-8JWm-wo0lE0C

Kaufmann, W. A. (1959). *The faith of a heretic*. Princeton, NJ: Princeton University Press.

Lamont, C. (1965). Mistaken attitudes toward death. *The Journal of Philosophy*, *62*(2), 29–36. https://doi.org/10.2307/2022993

May, T. (2014). *Death*. Routledge. Retrieved from https://market.android.com/details?id=book-K5DCBQAAQBAJ

Minerva, F., & Rorheim, A. (2017, August 8). What are the ethical consequences of immortality technology? Retrieved February 15, 2018, from https://aeon.co/ideas/what-are-the-ethical-consequences-of-immortality-technology

Nussbaum, M. C. (1994). *The therapy of desire: Theory and practice in Hellenistic ethics*. Princeton, NJ: Princeton University Press.

Parfit, D. (1984). *Reasons and persons.* Oxford: Oxford University Press. Retrieved from https://market.android.com/details?id=book-ulhHdvbDRUkC

Scheffler, S. (2013). *Death and the afterlife.* New York: Oxford University Press. Retrieved from https://market.android.com/details?id=book-yVh_AAAAQBAJ

Singer, P. (1987). Life's uncertain voyage. In J. J. C. Smart, P. Pettit, R. Sylvan, & J. Norman (Eds.), *Metaphysics and morality: Essays in honour of J.J.C. Smart.* Blackwell. Retrieved from https://philpapers.org/rec/SINLUV

The President's Council on Bioethics. (2003). *Age retardation: Scientific possibilities and moral challenges.* Retrieved from https://bioethicsarchive.georgetown.edu/pcbe/background/age_retardation.html

Williams, B. (1973). The Makropulos case: Reflections on the tedium of immortality. In B. Williams (Ed.), *Problems of the self* (pp. 82–100). Cambridge: Cambridge University Press. https://doi.org/10.1017/CBO9780511621253.008

Williams, B. (1976). *Problems of the self: Philosophical papers 1956–1972.* Cambridge University Press. Retrieved from https://market.android.com/details?id=book-RwWEgc-m0U0C

PART III

Alternative Uses of Cryonics

INTRODUCTION

In Part I of this book, we focused on the pros and cons of cryonics as a potential means of giving legally dead people another chance at life. We saw how cryonics holds up to various common arguments against similar technologies that society has dealt with through the years, and we identified some novel issues raised by cryonics in particular. In Part II, we discussed cryonics as a step towards radical and potentially indefinite life extension. We took a critical look at the badness of death in general, and saw how death can be considered harmful to anyone regardless of their age. Building on this conclusion, we then considered the pros and cons of immortality without ageing. We found that, although there are good reasons to believe that an indefinitely long life might be undesirable, there are equally good reasons to think that the current human lifespan is too short, and that we should at least try to extend it.

In this third and final part of the book, we will focus on the potential use of cryonics as a means of bypassing moral conflicts about two highly divisive medical practices in our society, namely euthanasia and abortion. The profound ethical disagreements about each of these practices arise from the fact that they both cause the death of a human being.

In the case of euthanasia, the human being in question is a person who is experiencing prolonged, incurable, and unbearable pain, and whose chances of recovery or relief are vanishingly small. People in these situations often find that the value of their life is negative, and that they would prefer to die rather than continue to suffer.

In the case of abortion, the human being in question is an embryo or a foetus developing in the womb of a woman who is either unable or unwilling to bring that human being into the world, and hence seeks to abort it. Her reasons for doing so might include wanting to spare the prospective child a life in suboptimal conditions, either because she would be unable to provide adequate care herself or because the foetus or embryo is affected by a medical condition that would cause severe disabilities; or she might simply not want to raise a child at that particular time, or at all, and instead devote her time to other pursuits.

I will argue that cryopreservation could provide a good alternative to both euthanasia and abortion, as it would achieve the desired result in each case, yet eliminate the necessary death of a human being by replacing an irreversible act with a reversible one. Cryonics would thus offer the advantage of reversing the decision to euthanize or abort, if and when circumstances are more favourable. In particular, it would allow a person who would choose *cryothanasia* (instead of euthanasia) to come back to life in case an effective therapy for their condition were to be found in the future, and it would allow a woman who chose to have a *cryosuspension of pregnancy* (instead of a termination of pregnancy) to reimplant the foetus if the circumstances that discouraged her from continuing the pregnancy were to improve. In the next two chapters, we will explore arguments in support of the thesis that cryothanasia would be a better option than euthanasia, and that cryosuspension of pregnancy would be a better option than abortion.

CHAPTER 5

Cryothanasia

Abstract Most objections to euthanasia are based on the moral principle that killing an innocent person is wrong. This principle also applies to cases wherein people ask (for help) to die in order to avoid unbearable, intractable, and incurable pain. It has been suggested that such patients could be offered an alternative in which they are cryosuspended immediately after their (legal) death has been medically induced. Such "cryothanasia" would allow them to be stored indefinitely with a non-negligible chance of being revived in a more medically advanced future. Since cryonics ultimately seeks to preserve and extend lifespan, these cryothanasia patients would, in effect, be choosing to die in order to (hopefully) live longer in the future. This chapter argues that classical objections to euthanasia, based on the principle that it is always morally wrong to kill an innocent person, cannot be used to oppose cryothanasia.

Keywords Euthanasia • Cryothanasia • Death • Cryonics • Assisted suicide

Euthanasia is any medical procedure wherein the explicit goal is to mercifully and painlessly end a patient's life in order to avoid unbearable suffering brought on by a (usually terminal) medical condition, upon a patient's request. The term is normally divided into two distinct types of cases. In the case of *active euthanasia*, a physician carries out the lethal procedure at the patient's request. In *passive euthanasia*, medical treatment is withheld or withdrawn from a patient, upon the patient's request, with the *intention* of shortening the patient's life (Giubilini, 2013); When referring to "euthanasia", broadly understood, I will also include the practice of *assisted suicide*, wherein the patient carries out the lethal procedure on themselves, albeit with the help and supervision of a physician who provides the necessary medical assistance (and, of course, always at the patient's request).

As of 2018, active euthanasia is only legal in the Netherlands, Belgium, Luxembourg, Canada, and Colombia, whereas assisted suicide is legal in Germany, Switzerland, Japan, and seven US states (Washington, Oregon, California, Colorado, Montana, Vermont, and Washington, DC). Patients in these countries who experience prolonged and incurable suffering can request to hasten their own death in a controlled manner. Such requests are only granted as a last resort after the patient has already gone through all standard treatments for their condition, and there are regional variations around what conditions are seen as sufficient grounds for euthanasia.

The goal of euthanasia is to stop unbearable and intractable suffering, hence the colloquialism "putting someone out of their misery". Patients who request euthanasia do so out of an overwhelming desire to be relieved of severe pain—whether physical, psychological, or both—and after concluding that currently available therapies are inadequate. Realizing that continuing their life will almost certainly imply never-ending misery, these patients then opt for death as the only available alternative. The fact that ending their plight at that point necessarily includes ending their life is, by and large, an unfortunate side effect: if it were possible to end the suffering (either physical or psychological) without ending the life, of course euthanasia would not be a rational or an ethical option. So, if there were some way to relieve these patients of their suffering without simultaneously causing them to die, we can reasonably assume that the vast majority of them would consider it a preferable option. It is easy enough to imagine such options coming into existence at some point in the future, and we can draw on historical examples for inspiration. Advancements in medical science have helped ensure a decent quality of life for patients with conditions that, until recently, were seen as more or less synonymous with misery. Minor injuries and infections were often fatal before the advent of modern evidence-based medicine in the late nineteenth century, and those lucky enough to survive were often left with lifelong complications and chronic pain; migraines, cluster headaches, and other neurological pain disorders only became treatable in the early twentieth century; and there were practically no effective treatments for severe psychiatric disorders before 1950. Hence, it is reasonable to assume that among today's most painful and intractable medical conditions, at least some will become treatable in the future. In other words, at least some of today's euthanasia patients would find their prospects improved if only they could stick around for a few more years or decades. Of course, this is old news to those who opt for euthanasia. When a patient's suffering becomes so unbearable that they consciously decide to end their life, they simultaneously give up on waiting for a better alternative.

In a paper I co-authored with Anders Sandberg, we discussed what kind of alternative could technology provide to people who find themselves in such a difficult situation (Minerva & Sandberg, 2017). The answer we came up with was *cryothanasia*: a hypothetical future procedure with the explicit goal of painlessly pausing (rather than ending) the life of a patient who would otherwise satisfy the conditions for receiving euthanasia, in the hope of someday being able to resume their life, cure their condition, and enable them to live a full life.

By undergoing cryothanasia, a patient practically leaves open the possibility that there will be at some point an alternative to death available to him or her, and therefore the intention behind cryothanasia is not that of ending one's life in order to avoid suffering, but that of postponing the end of one's life until suffering could be avoided without procuring death.

The terminological choice of the word *cryothanasia* means, from a merely linguistic point of view, the loss of the *eu-* ("good") component of the concept of *euthanasia*, which refers to a kind of death that is good for the patient. Now, the terminological choice might not be seen as very appropriate, given that I have specified that the patient undergoing cryothanasia does satisfy the conditions for receiving euthanasia, and therefore the procedure is meant to be good (*eu-*) for the patient by terminating a condition of untreatable prolonged unbearable suffering. However, I want to stick with the term *cryothanasia* (that I chose together with Anders Sandberg) because it effectively conveys another important meaning, namely that the outcome is not death, as the term *-thanasia* implies, but something different, which is achieved through cryonics rather than by an act of merciful killing. While we could therefore consider other terms such as "cryodeath", "pseudodeath", or "cryocide", as suggested by Ole Martin Moen (2015), I prefer to use the term *cryothanasia* because its assonance with euthanasia reminds us that it is a procedure that is always done in the interest of the patient undergoing it.

Apart from the general implications of cryonics covered in the rest of this book, what makes cryothanasia especially interesting from an ethical standpoint is that it appears to circumvent many of the most frequently cited arguments against euthanasia. Perhaps most notably, if one believes that the fundamental goal of any medical practice should be to improve health and extend life, then euthanasia clearly falls outside the domain of morally acceptable medicine (since it aims at killing the patient). But since the goal of cryonics is to greatly improve health and lifespan, and cryothanasia would improve its chances of success, then it becomes very difficult to argue that cryothanasia is impermissible on the same grounds as traditional euthanasia (Shaw, 2009).

Cryothanasia is ethically different from euthanasia in many other important respects. This leaves objections to euthanasia largely inapplicable against cryothanasia. Hence, valid objections to cryothanasia must rely on some set of reasons that do not concern traditional euthanasia, which I will explore in the latter part of this chapter.

We can now turn to an ethical analysis of cryothanasia. I will do this in two ways. First, in light of the characterization of cryothanasia and of its differences with euthanasia I have offered above, I will rely on the standard objections commonly raised against euthanasia. As we shall see in the next section, the differences between cryothanasia and (certain types of) euthanasia which have emerged in this section imply that even if we assume—for the sake of argument—that the standard objections to (certain types of) euthanasia are valid, they do not imply the moral wrongness or badness of cryothanasia, and actually imply that cryothanasia is morally preferable to euthanasia. Second, I will consider a few other possible objections to cryothanasia that are independent from the standard objections to euthanasia, and I will show that even this second group of objections is not strong enough to make cryothanasia a morally impermissible enterprise.

Objections to Euthanasia Applied to Cryothanasia

We will now review the most common arguments against euthanasia and the degree to which they also apply to cryothanasia. As we will see, cryothanasia tends to escape such arguments.[1]

Deontological

By "deontological" objections to euthanasia I mean objections that appeal not to the badness of the outcome in itself, but to the intrinsic wrongness of the act itself or of the reasons, intentions, or motivation with which the act is carried out. There are two main types of deontological objections to euthanasia: one based on the moral relevance attributed to the distinction between acts and omissions and one based on the moral relevance attributed to the distinction between intention and foresight, which in turn is grounded on the so-called doctrine of double effect (DDE). Let us consider these two types of objections in order to see whether they apply to cryothanasia.

Let us start from a moral distinction that many people draw between active and passive euthanasia. It is held by some that active euthanasia is immoral because it actively causes the death of a person, as opposed to passively allowing a person to die. This objection is based on the belief that acts (or actions) carry more moral weight than omissions. In the context we are examining, this means that causing the death of a person is morally unacceptable, while failing to prevent that same person from dying naturally is morally acceptable—or at least less wrong than active euthanasia.

Assuming that the distinction between actions and omissions does carry a moral weight such that passive euthanasia is indeed morally different from active euthanasia, the reason why the act in question is considered bad or wrong is that its outcome, namely the death of a patient, is considered necessarily something morally bad. But the outcome is different in the case of cryothanasia and of euthanasia. Whereas euthanasia causes certain death, cryothanasia only results in a possibility of death. Surely, the ideal alternative to certain death would be an act that restores health and allows the patient to keep living right away; but in the absence of such alternatives, reducing a certain death to an uncertain chance of death is an improvement, if only a small one, and indeed something morally preferable on any plausible conception of morality (Shaw, 2009).

Another common objection to euthanasia is based on the DDE. In short, DDE states that if a certain action has two outcomes, one of which is good (e.g. the end of suffering) and the other one is bad (e.g. death), then the action is only permissible if the bad outcome is not intentionally used as a means for the good outcome, and the badness of one outcome is proportionate to the goodness of the other outcome, and if the action itself is not intrinsically morally wrong. The DDE belongs to the Thomistic tradition and is typically embraced by the Catholic Church. In the case of euthanasia, even though the intended goal of relieving suffering is beneficial, the act of killing is morally bad. Hence, doctors operating under DDE normally do not condone euthanasia (although they may condone what is sometimes referred to as "palliative sedation", i.e. giving patients doses of sedatives and/or analgesics that would foreseeably hasten death, as the act of administering such drugs to relieve pain is morally neutral and the intention is that of relieving pain, not killing the patient).

Cryothanasia more closely meets the requirements of the DDE criteria. The action, a hopefully reversible and (hopefully temporary) pausing of life, is arguably not as wrong as the killing of a person even on moral views that consider killing an innocent man always morally wrong. It has two goals, relieving pain and extending life, each of which is beneficial. One could argue that the procedure itself is impermissible because it entails a risk of death, which would be a bad outcome, and the intended benefit does not outweigh the risk of cessation of life. However, even if that was the case, death or risk of death is an unintentional side effect of the procedure, and not—as in the case of euthanasia—the intended outcome. Granted, one could argue that (hopefully temporary) pausing of life is somehow inherently bad, but it is not clear how one would convincingly explain where this inherent badness stems from; it might be wrong to artificially keep a person unconscious because that would temporarily deprive that person of a potentially meaningful life, but it is not clear how this would be different from normal general anaesthesia.

Thus, we can conclude that the standard deontological objections against euthanasia do not apply to cryothanasia.

Faith-Based

Western religions tend to judge euthanasia on a basis similar to that of DDE: they consider the relief from suffering a good thing or at least not a bad thing[2] unless it is achieved by killing the patient, which, if the patient is an innocent person (which he or she will likely be), is always morally wrong. In addition to how cryothanasia avoids DDE, as stated above, it is possible to argue that cryothanasia is not about relieving suffering per se, but rather about avoiding death. Although many religious and secular views do place some positive value on death (whether instrumentally, for instance, as a means of getting to the afterlife, or intrinsically) they tend to view premature death as largely negative, and most prohibit seeking death in all but extreme cases (such as martyrdom).

Moreover, even if one believes there is something valuable to be learned by enduring terminal suffering, it would nevertheless seem that a person would have the time to learn much more if they were able to live a full life, especially after a brush with death.

Another religiously based objection argues that life is a sacred gift from God, and that humans have a duty to respect and protect life at all cost. In this view, euthanasia is a sacrilege no matter how badly one is suffering. Cryothanasia, however, seeks to preserve and ultimately prolong life in cases where the only alternative is suffering followed by certain death. Hence, one could argue that cryothanasia not only respects the sanctity of life, but actively demonstrates a profound personal dedication to it. Or at the very least, cryothanasia is not significantly different from all other medical interventions that seek to prolong life, which are normally not condemned by religions (unless the treatment is considered futile) and which indeed are often actively supported by religious doctrines, such as in the case of life-prolonging treatments on a terminally ill patient. There is, of course, still a risk that revival will not succeed, in which case cryothanasia amounts not to a case of euthanasia, but rather a case of inadvertent death during a medical procedure intended to save the patient. The risk would thus be comparable to the risks associated with complex life-saving surgeries or experimental cancer treatments, which are generally considered acceptable under mainstream religious views (Bridge, 2015).

Principles of Medical Ethics

One of the most common objections to euthanasia grounded in medical ethics appeals to the Hippocratic Oath, which states that the purpose of medical practice must always be to heal, never to kill. Although classical versions of the oath specifically prohibit euthanasia, it is mostly excluded from modern official medical oaths,[3] which tend to be more vague. Many supporters of the Hippocratic Oath would argue that since euthanasia is generally viewed as a personal choice, it should thus be a personal (or at least non-medical) responsibility to perform it. Many also argue that even in countries where euthanasia is legal in some form or another, physicians should and in fact do retain the right to refuse requests to perform or assist in euthanasia, in the form of "conscientious objection". In fact, all the legislations in the world that allow some form of medical assistance to dying include conscience clauses that recognize doctors' rights not to take part in this form of assistance.

The degree to which cryothanasia is consistent with the prescriptions of the Hippocratic Oath depends in part on how one defines death. Cryothanasia would certainly involve bringing about *clinical* death, but it would not cause *information-theoretic* death (we defined this account of death in Chap. 1), and would even seek to actively prevent it. Being ancient in origin, the oath of course does not specify any particular definition of death, so we would need to determine which conception of death is more morally relevant in the case of cryothanasia, with regard to consistency with the aims and scope of medicine. It seems plausible that, consistently with the prescription of the Hippocratic Oath, the kind of death which involves the permanent loss of a human life is what medicine should try to avoid or prevent as much as possible. Thus, once the option of cryothanasia is introduced, it is the information-theoretic death, and not clinical death, that medicine should try to avoid or prevent. We can thus easily imagine a modern physician who, without any internal conflict, swears by a strong form of the Hippocratic Oath without having to oppose cryothanasia.

Besides, even if we assume a clinical definition of death as the morally relevant one, it appears there are still cases in which cryothanasia would adhere to the Hippocratic Oath. Modern versions of medical oaths often contain an obligation to apply "all measures required" for the benefit of sick patients, avoiding both "overtreatment and therapeutic nihilism". In scenarios where cryothanasia truly is the last chance of survival for a dying patient, and where the overarching goal is to ensure their recovery to the best of one's ability, it is difficult to see how the procedure would not qualify as a required measure for the patient's benefit. At the very least, it would seem no less acceptable than other last-ditch efforts to save patients from certain death, including various experimental treatments with low odds of success. Of course, one could also argue that since cryothanasia is both extremely invasive (involving total replacement of all bodily fluids, among other measures) and very expensive, it would amount to overtreatment even when proposed as a last resort; conversely, however, one could argue that refusing cryothanasia in such cases would amount to a weak form of therapeutic nihilism, that is, a failure to do whatever is in the power of medicine to prevent a patient's death.

Another objection based specifically on medical ethics is the one according to which when patients ask for euthanasia in order to be relieved of grave physical or psychological suffering, we should in reality provide them with better palliative care rather than hasten their deaths. In this view, requests for euthanasia indicate a failure on behalf of palliative care in offering adequate relief therapies to suffering patients, and euthanasia would be unnecessary if adequate therapies were available.

But while euthanasia, palliative care, and cryothanasia do share the goal of relieving suffering, cryothanasia also has a unique and second goal which neither palliative care nor euthanasia have and which is supported by principles of medical ethics, namely that of benefitting the patient by extending her lifespan. Patients who request cryothanasia would not be doing so only out of a lack of palliative care, but also out of a desire to live beyond their current prognosis—albeit in a more remote future. It should also be noted that palliative care is compatible with cryothanasia to some extent. Indeed, any palliative care treatment that do not significantly damage the patients' chances of a successful cryopreservation and revival would be advisable in the days and weeks leading up to cryothanasia.

Some also argue that euthanasia gives doctors a power over life and death that is by itself improper, and that further unbalances the already skewed power relationship between doctors and patients.

While it may well be the case that doctors should not exercise undue power over patients, it is not at all clear that euthanasia, and much less cryothanasia, would be examples of such power. Doctors normally have great power in choosing what diagnoses and treatments to give each patient. Patients, in turn, trust that doctors will always have the patient's best interests in mind when using their expert judgement in making treatment recommendations, which patients are then free to either accept or refuse. With euthanasia and cryothanasia, however, this relationship is turned upside down. Each patient personally decides on a specific intervention, and then suggests the intervention to their doctor, who can then choose to fulfil the request personally or delegate the matter to someone else, in case of a conscientious objection. Therefore, the issue of doctor's undue power does not arise in the case of euthanasia or cryothanasia.

Weirdness and Repugnance

As we saw in the first chapter, when new technologies are introduced to the public, they are often perceived as "weird" or "unnatural" or "yucky". There are two senses in which cryonics and cryothanasia certainly come across as weird to many of us. First, they are weird in the sense of being unusual, that is, options that we have never had to deal with before and about which therefore many of us do not have strong intuitions, either in favour or against. Second, they are weird in the sense that they radically change our paradigmatic views of what it means to be dead and to be alive, by introducing a third option that shakes many of our fundamental ethical and religious views about the meaning of life, death, and (im)mortality; and, as is the case with most novelties of this kind, people tend to experience an intuitive and emotive rejection of the option in question not so much by virtue of its distinguishing aspects, but because of the novelty itself. This reaction can be explained by a negativity bias, that is, the tendency to see the negative aspects rather than the positive ones in any situation, which generally characterizes conservatives' approach (Hibbing, Smith, & Alford, 2014), and a status quo bias, that is, the tendency to prefer the options that preserve the status quo over the options that alter it.[4]

The first obvious reply to this consideration is that it is obviously fallacious: that something is or seems weird does not constitute a good reason for considering it morally wrong, at least if we believe that the morality of an action or of a practice is at least in part independent of our psychology. The second, and related, reply is that many medical practices which are today commonly accepted and indeed sometimes considered morally good looked very weird when they were first introduced; two obvious examples are heart transplant and in vitro fertilization. Therefore, the weirdness of cryothanasia does not mean that this practice also will not one day be accepted and indeed considered by many as morally good. Finally, practices which are considered weird from outside certain specific groups, such are circumcision and refusal of blood transfusion, have become part of medical practice despite not aligning with mainstream medical ethics, and there is no reason to think that the same ethical reasons that motivated the introduction of these practices would not justify performing cryothanasia as well.

Unlikelihood and Futility

Another relatively obvious objection to cryothanasia is that it is unlikely that cryonics will ever work. Therefore asking to be cryopreserved and to lose some certain lifetime in the hope to gain an unlikely future extension of life is a risk not worth taking, and people should not be offered the option to choose to take risks not worth taking. This argument applies differently in the case of cryonics performed upon natural death and in the case of cryothanasia. The risk in the former case is always worth taking regardless of the likelihood of the success of cryonics, because the person who naturally dies has nothing to lose and everything to gain; the risk in the case of cryothanasia does not seem to be worth taking, at least if the person who chooses cryothanasia had quite a long life ahead of him or her before natural death, and if the foreseeable quality of this lifespan were good enough.

However, three considerations can be put forward to respond to this type of objection. First, it is very difficult to estimate the actual likelihood of success of cryonics and cryothanasia; until the point at which success is very close, it would remain very difficult not only to predict this success but to estimate its actual likelihood—therefore, whether the success of cryothanasia is really so unlikely is something we cannot simply take for granted. Second, even if the chances of success are very low, cryothanasia could still represent a risk worth taking for those for whom natural death would be close anyway and/or the expected quality of the expected remaining lifespan is sufficiently low. Third, even if the risk were not worth taking, it could be argued that a principle of liberty should prevail and that, therefore, people should be free to take the risks they autonomously choose to take. The important aspect, in this case, is that the choice of the person who opts for cryothanasia should be really autonomous; that is, there should be no form of coercion, including psychological pressure, from other people to choose cryonics. The real ethical problem would then become not one about the unlikelihood of success of cryothanasia, but one about the possibility of coercion. However, this is not an exclusive problem of cryothanasia: it applies also to euthanasia and indeed to any kind of medical intervention. There is no reason to think that it should be more of a problem in the case of cryothanasia than in the case of any other medical intervention.

These kinds of responses, and the first one in particular, should also address a related concern, namely the concern about the futility of cryonics and cryothanasia as medical treatments. They would be equivalent to any other kind of futile medical treatment that is normally considered not morally permissible. However, are cryothanasia and cryonics really futile? Futility is defined as "a situation where the evidence shows no significant likelihood of the treatment conferring a significant benefit" (Minerva & Sandberg, 2017). But in the case of cryonics the evidence is simply missing, either in one sense or in the other. Therefore, the treatment is not futile, but experimental. Being so, the lack of evidence about likelihood of success, far from being a reason for not undergoing the treatment, is a reason for welcoming people who voluntarily choose cryothanasia, as they would give scientists the opportunity to try an experimental treatment and gather evidence about the likelihood of its success—indeed, experimenting cryothanasia on volunteers will increase, in an ethical way, the chances that cryothanasia will at some point be successful. Thus, there are reasons to think that cryothanasia should be protected under so-called right to try laws, at least in the case of terminally ill patients whose only alternative is an immediate death, and for whom therefore a treatment with even a slightly more than zero chance of success (like cryothanasia) would be in their best interest (ibid.).

Resource Use

Another objection holds that cryothanasia and cryonics in general will use up resources that could be destined to more urgent needs than that of extending someone's life indefinitely. This objection tends to come up whenever a new medical option promises outcomes beyond the scope of traditional medical practice (think, again, of the case of in vitro fertilization). However, to keep a person cryopreserved for a long time is relatively cheap, as it only requires fortnightly liquid nitrogen refills. The start of the prevention process itself would certainly be expensive: however, apart from the fact that it would be paid by the patient themselves (I am not going to discuss here whether or to what extent cryonics should be paid for by a public health system), it is worth noting that cryothanasia, especially if it becomes relatively popular, would free up many resources currently used for end-of-life care. Therefore, not only does the resources objection not provide an argument against cryothanasia, but actually it could be used to put forward a counterargument: cryothanasia might indeed reduce the need for medical resources.

It seems that the objections we have examined here—including those derived from the standard objections against euthanasia—either do not apply or are not strong enough to outweigh the reasons in favour of cryothanasia. Cryothanasia seems more strongly ethically justified in the case of terminally ill patients. Now, in one sense, every one of us is terminally ill, because ageing itself is a terminal condition. However, it seems that for those who have a likely long lifespan ahead, cryothanasia is not worth the risk, and it is not a rational choice, even if one of the possible (though we do not know how probable) outcomes is the possibility of living an indefinitely long life once revived. This does not mean that there are sufficiently strong reasons to deny those who would nonetheless autonomously choose cryothanasia in such circumstances the possibility of carrying out their plans (ibid.). But there might be good reasons to at least try to dissuade these individuals from undergoing cryonics. In the case of terminally ill patients with a very poor prognosis and a short lifespan ahead, however, not only would there be no good reasons to prevent or dissuade them from undergoing cryothanasia, but indeed there would be good reasons to encourage people to choose cryothanasia as the option that is in their best interest.

Notes

1. For an overview of the main ethical issues raised by euthanasia in the medical context, see Keown (1997).
2. It should be noted that some religious people nevertheless consider suffering as a positive thing in certain contexts; see Paul (1984).
3. Orr, Pang, Pellegrino, and Siegler (1997) notes that only 14% of official medical oaths specifically prohibited euthanasia as of 1993.
4. For a philosophical discussion of status quo bias, see, for example, Bostrom and Ord (2006).

References

Bostrom, N., & Ord, T. (2006). The reversal test: Eliminating status quo bias in applied ethics. *Ethics, 116*(4), 656–679. Retrieved from https://www.ncbi.nlm.nih.gov/pubmed/17039628

Bridge, S. W. (2015). Why a religious person can choose cryonics. In A. De Wolf & S. W. Bridge (Eds.), *Preserving minds, saving lives: The best cryonics writings from the Alcor Life Extension Foundation*. Alcor Life Extension Foundation. Retrieved from https://market.android.com/details?id=book-6QgvjgEACAAJ

Giubilini, A. (2013). Euthanasia: What is the genuine problem? *The International Journal of Applied Philosophy, 27*(1), 35–46. Retrieved from https://www.pdcnet.org/ijap/content/ijap_2013_0027_0001_0035_0046

Hibbing, J. R., Smith, K. B., & Alford, J. R. (2014). Differences in negativity bias underlie variations in political ideology. *The Behavioral and Brain Sciences, 37*(3), 297–307. https://doi.org/10.1017/S0140525X13001192

Keown, J. (1997). *Euthanasia examined: Ethical, clinical and legal perspectives*. Cambridge: Cambridge University Press.

Minerva, F., & Sandberg, A. (2017). Euthanasia and cryothanasia. *Bioethics, 31*(7), 526–533. https://doi.org/10.1111/bioe.12368

Moen, O. M. (2015). The case for cryonics. *Journal of Medical Ethics, 41*(8), 677–681. https://doi.org/10.1136/medethics-2015-102715

Orr, R. D., Pang, N., Pellegrino, E. D., & Siegler, M. (1997). Use of the Hippocratic Oath: A review of twentieth century practice and a content analysis of oaths administered in medical schools in the U.S. and Canada in 1993. *The Journal of Clinical Ethics, 8*(4), 377–388. Retrieved from https://www.ncbi.nlm.nih.gov/pubmed/9503088

Paul, P. J., II. (1984). *Salvifici doloris*. Ediciones Paulinas. Retrieved from http://catholicsociety.com/documents/john_paul_ii_letters/Salvifici_doloris.pdf

Shaw, D. (2009). Cryoethics: Seeking life after death. *Bioethics, 23*(9), 515–521. https://doi.org/10.1111/j.1467-8519.2009.01760.x

CHAPTER 6

Cryosuspension of Pregnancy

Abstract A woman who finds herself pregnant against her plans (e.g. because contraceptives have failed) has only two available options: continuing or terminating the pregnancy. Continuing the pregnancy may not be an option due to lack of economic resources, possible birth defects, or other life plans; yet terminating the pregnancy may be considered immoral by the woman, the society in which she lives, or both, and the decision may thus be a source of great distress. This chapter suggests that a hypothetical future technology aimed at extracting and cryopreserving foetuses could provide unwillingly pregnant women with an alternative to the dilemma of either continuing or terminating the pregnancy. Cryopreserved foetuses could be reimplanted at a later time, when circumstances are more favourable to continuing a pregnancy and raising a child.

Keywords Abortion • Termination of pregnancy • Cryosuspension of pregnancy • Cryonics • Ectogenesis

In the previous chapter, we considered how cryothanasia—the cryopreservation of an individual who would otherwise choose to be euthanized or to commit suicide—could be used as a practical option to bypass issues caused by disagreement about the moral permissibility of choosing death. We saw how, rather than mercifully kill someone experiencing unbearable and incurable suffering and asking for euthanasia, we could instead cryosuspend them in the hope that science will someday find a way to eliminate the cause of their suffering (and, obviously, to revive them).

Of course, euthanasia is not the only divisive medical practice in our societies. Abortion is perhaps even more controversial than euthanasia, as it involves the killing of a human being who, for obvious reasons, cannot give consent to the procedure. So, whereas in the case of euthanasia, the request to be killed comes from an autonomous and rational person, this is not the case with abortion, where the killing is not justified on the basis of the fact that the embryo or foetus is asking to be aborted. Arguments justifying abortion are more complex than those based on the simple right to autonomy involved in euthanasia, and the whole debate on abortion is therefore more complex and more heated.

Even though abortion is legal in many more countries than euthanasia is ("The World's Abortion Laws Map", 2014), the debate about the moral permissibility of abortion is as lively today as it was 50 years ago. People hold deeply felt opinions about the morality of abortion, and such opinions seem to be rooted in non-negotiable ethical views. The ongoing disagreement about the moral value of embryos and foetuses, and about whether to prioritize women's bodily autonomy (the pro-choice view) or foetuses' (alleged) right to life (the pro-life view), seems practically unsolvable. Even though clear arguments have been developed on both sides of the debate, it seems that it is very difficult to move people who hold pro-choice views to the pro-life side, and vice versa. But even among people who share pro-choice or pro-life views, there are disagreements surrounding the reasons why abortion is moral or immoral, and regarding what exceptions to the rule, if any, should be allowed within a given moral framework.

For instance, some people on the pro-choice side argue that abortion is permissible only up to a certain stage of the pregnancy. Similarly, some people on the pro-life side would consider abortion permissible if the pregnancy is the result of sexual assault, or the pregnancy somehow threatens the pregnant woman's health or life.

From the pro-life side, abortion is considered immoral because it involves the death of the embryo/foetus. Causing the death of an embryo/foetus is considered wrong either because it violates the principle that one should never kill an innocent human being or because it would be in the best interest of the foetus to continue to develop and have a life. From a pro-choice perspective, meanwhile, abortion is permissible either because embryos and foetuses are not yet fully human in a morally relevant sense (hence it is morally permissible to kill them) or because a woman's autonomy over her own body cannot be outweighed by the right to life of any individual who depends on the woman to survive, regardless of their moral status.

In this chapter, I will consider how cryosuspension of embryos and foetuses could serve as an alternative to the termination of an unwanted pregnancy and thereby help bypass moral disagreements about abortion.

In a paper co-authored with Anders Sandberg, we considered the positive impact that could result from developing a new technology aimed at cryosuspending foetuses (Minerva & Sandberg, 2015). Such technology could prove helpful in emancipating women from biological constraints related not only to unwanted pregnancy, but also to unwanted decrease in fertility due to ageing.

Giving Pregnant Women Another Option

Medicalizing the reproductive process has brought about radical changes in the way we approach parenthood. For instance, as we saw in the first chapter, in vitro fertilization (IVF) and embryo cryopreservation (EC) have become common procedures, and hundreds of thousands of children have born thanks to these technologies. IVF allows embryos to be created in the lab, after which they can be stored almost indefinitely using EC. Most cryopreserved embryos are later implanted when it is most convenient for the mother and then let develop into foetuses, others remain at the embryonic stage and may be used for research purposes, and the rest are eventually destroyed. Contraceptives, too, have had a profound impact on family planning and procreation, allowing women to choose whether, when, and with whom to have children.

However, when a woman finds herself pregnant against her plans, for instance, because contraceptives have failed, she has only two available options: continuing or terminating the pregnancy. In the future, technology could provide unwillingly pregnant women with a third option, namely cryosuspension of pregnancy—thus removing the need to choose between having a child they either cannot have or do not want to have at that particular moment, and undergoing an abortion they either do not want to have or cannot have for moral, legal, or personal reasons.

No attempt has been made (yet) to extract an embryo or a foetus already implanted in a uterus in order to cryopreserve it. If the extraction and the cryopreservation were successfully performed, the embryo/foetus could remain cryopreserved for years, and be reimplanted at a time where a woman feels ready to continue the pregnancy and raise the child.

At a purely theoretical level, there is no reason to assume that the cryopreservation of an embryo or foetus extracted from the womb would not work (Pavone, Innes, Hirshfeld-Cytron, Kazer, & Zhang, 2011), but the process of extracting a human being of few cells, or after a few weeks of development could be particularly difficult. To begin with, the extraction of the embryo/foetus from the uterus could damage the embryo/foetus, the uterus, or both. This part seems to be the one presenting the biggest technical difficulties: embryos (and, to a lesser extent, foetuses) are extremely small and delicate, and extracting them from a womb without damaging or accidentally killing them would not be easy. But new and sophisticated techniques for performing surgeries on extremely small and fragile tissues are under constant development, so we should not exclude the possibility that such technical obstacles will someday be overcome.

Alternatively, the same result could be obtained by inducing the preterm birth of a foetus of viable age (around 23 weeks) and then proceeding with the cryopreservation. This option would remove some of the risks associated with removing an extremely small and fragile organism from the womb, but it would present the inconvenience of requiring women to continue the pregnancy for several weeks, which could cause psychological and/or physical discomfort to at least some of them.

Even though such technical difficulties should not be underestimated, they do not constitute a valid reason for discarding cryosuspension of pregnancy altogether. As I have discussed in previous chapters, it is crucial to assess the moral permissibility and expected utility of any given project well before technical feasibility is achieved. If there are not enough good reasons to try to develop a technology, or if there are enough good reasons to avoid the outcome that the technology hopes to achieve, then resources should not be invested in the project at all. This is why ethical analysis must always be one step ahead of the technology itself, and why it is important to discuss the implications and overall worthiness of potential future technologies as well as possible future developments of existing technologies.

In the following pages, we will explore the reasons why both abortion and continuation of an unwanted pregnancy can be suboptimal solutions under some circumstances, and how cryosuspension may offer a way out of such dilemmas.

Would Objections to Abortion Apply to Cryosuspension of Pregnancy?

According to philosopher Don Marquis, abortion is not morally permissible because it deprives embryos and foetuses of their future, which—given that embryos and foetuses have their own future ahead of them in the same way as the rest of us have our own futures ahead—is as valuable as anyone else's future (Marquis, 1989). Just as it is wrong to deprive adults of their future life, even though they do not know what the future holds in store for them, it is also wrong to deprive embryos and foetuses of "a future like ours", in Marquis' own words.

As we saw in Chap. 3, Thomas Nagel argued that one is harmed by death even if one no longer exists after one is dead, because the harm of death consists in being deprived of the counterfactual life one would otherwise have continued to enjoy.

These two arguments can be tied together to form an argument against abortion, based on a kind of harm that abortion inflicts on the embryo or foetus. In the same way that we can imagine a counterfactual life for a person who died in their 30s, and say that their death was bad because it deprived them of a future life, we can also imagine a counterfactual life for the embryo or foetus being aborted, and say that their abortion was bad because it deprived *them* of a future life.

However, one could reply by pointing out that the two types of future at stake—the one of a foetus and the one of an actual person—are not morally equivalent, in the sense that being deprived of one is not as bad of being deprived of the other. In this view, before a person becomes minimally self-aware, their counterfactual life is not morally relevant in the same way that the counterfactual life of someone who has not been and never will be conceived is morally (ir)relevant. This line of argument boils down to a crucial difference between attributing a counterfactual life to an entity that is not developed enough to appreciate the fact they are and will be alive, and attributing a counterfactual life to an actual person such as myself: while my counterfactual life has value because my current self is capable of attributing value to it, the counterfactual life of an embryo or a foetus has no value because the embryo or foetus does not have the capacity to attribute any value to it (Giubilini, 2012).

Regardless of when during foetal development one thinks it would start being wrong or harmful to deprive a human being of their counterfactual future, it seems that cryosuspension of such individuals would not entail any such harm. The cryopreservation of this embryo or foetus would not deprive it of a future life, but merely postpone the beginning of their "biographical" life (as opposed to their biological life, which would nevertheless start at conception). So if one gets pregnant in 2020 and decides to extract the embryo or foetus from the womb and cryopreserve it, then insofar as the embryo will be reimplanted in the uterus at some point in the future, that embryo/foetus has not been deprived of its future life.

What Type of "Future Like Ours"?

One could object that, even if an embryo that is extracted, cryopreserved, and then reimplanted has not been deprived of its future altogether, it has nevertheless been deprived of the *particular* biographical life that it would have had, had it been allowed to develop "naturally" after its conception. In other words, it would have been deprived of "its" own future, which would have been replaced with a different future.

Thus, an embryo that was conceived in, say, 2020, but then cryopreserved and eventually reimplanted in 2040, would be deprived of the counterfactual life it would have lived if it had developed naturally and been born nine months after conception. Of course, the biographical life one will have if reimplanted and born 20 or more years after they were conceived will be different from that which they would have experienced had they been born 9 months after conception. For example, consider how a child born in the 1980s grew up without access to the internet or mobile phones. Given how each of those technologies ended up radically changing our lives, it would be false to say that the same embryo born in the 80s rather than today would have had the same biographical life. And, of course—besides the different kinds of technologies available to people growing up in different decades—there would be major differences in the kind of education, culture, and social environment to which the child would be exposed, all of which would greatly influence their life in a multitude of ways.

But even though it is obvious that the same individual born today versus 30 years ago would have a different biographical life—a different narrative, so to speak—it is far from clear whether the embryo would have an interest in being born at one time rather than the other; nor is it clear how they might be harmed by being born 10, 20, or 30 years after they were conceived. Just like embryos that are created in the lab and implanted at a later time are not harmed by the fact that they spent some years or decades as cryopreserved embryos, embryos and foetuses reimplanted after cryosuspension of pregnancy should not be considered harmed by either this process or the delay of coming into existence. Moreover, we need to remember that the realistic alternatives for these embryos and foetuses would have been either abortion or being brought into existence under suboptimal conditions, as their parents were not ready to take care of them.

Only if one could predict future events that would shorten the overall life of a future child, or negatively affect their overall well-being, there might be a reason to avoid the cryopreservation of the embryo. For instance, if one knew that an asteroid would impact the Earth and destroy human civilization in 2060, they would have good reasons to bring that future child into existence as early as possible, thereby giving it as many years of life as possible before certain death in 2060. But in the absence of information of this sort, it seems that cryopreservation of embryos/foetuses is not harmful, at least insofar as they are eventually implanted.

So, if one is against abortion for the reason that it deprives the embryo or foetus of a future like ours, then cryopreservation after removing it from the womb would be a better alternative to abortion.

Potentiality

Deprivation of a future life, however, is not the only reason why abortion is considered morally impermissible. According to a different objection, abortion is immoral because it causes the death of a potential person. Embryos and foetuses are potential persons, meaning that, in absence of adverse events, they can develop into people. While it is true that they are human beings in a biological sense (i.e. they are members of the species *Homo sapiens*) from conception, under certain views being a *person* is different from merely being human. On certain views, a person is an individual with self-awareness—the capacity to appreciate that they are alive—and an interest in continuing to live. Some humans, like indeed embryos and foetuses, do not meet such requirements. Given that embryos or foetuses do not have these capacities and hence do not meet the requirements for personhood, they are not persons in the same sense that an adult human being is a person. However, they surely have the potential to *become* a person like you and me, but there is disagreement about whether potential persons and actual persons should be attributed the same moral status. Let us briefly illustrate how the opposite argumentative lines go.

Let us start from the pro-choice view. If someone kills me (a fully developed adult person), they frustrate my preference to keep living, reach certain goals, develop certain skills, and so on. However, if someone had killed me when I was just an embryo, they would not have prevented me from achieving my goals, since I would not have had any at the time. Not all living beings (humans or not) have an interest in continuing to live, as such interest only develops in beings capable of attributing a value to their being alive, and to fulfilling their projects. As we saw in Chap. 3, for instance, Bernard Williams attributed to "categorical" desires the power to propel us into the future: if such desires never existed, or are depleted over a long life, then one has no interest in living. The odd implication of this approach is that newborns and people with severe cognitive impairments also have no interest in living, hence their death is not bad to them (Giubilini & Minerva, 2012).

According to the opposite view, being someone with the potential to become a person with an interest in continuing to exist is sufficient for being attributed a right to life. It does not matter if the criteria for calling someone a "person" are actually satisfied or not; if someone has the potential to become a person, they should be treated as such. An odd implication of this view is that, according to this argument, eggs and sperm should also be considered potential people, as they too develop into persons given the right set of circumstances. And yet, we do not think there is anything bad about the monthly expulsion of an unfertilized egg, even though, had they been fertilized, they could have become a person. Moreover, within two weeks following conception, an embryo can split itself into several twins, or even become a kind of tumour called "teratoma". So it is difficult to tell whether those embryonic cells are indeed a potential person, several potential people, or a potential tumour (Ford, 1991). Thus, when arguing against abortion on the basis that it is the killing of a potential person, one has to deal with the difficulties of drawing a line between what does and does not count as a potential person.

However, also in this case it does not matter, for the purpose of assessing the moral permissibility of cryosuspension, whether the argument from potentiality as applied to abortion is successful or not. Cryosuspension entails neither the loss of the embryo/foetus nor the loss of its potentiality, since, once revived, the embryo/foetus will continue to develop and realize its potential. Development of potentiality is delayed, but not prevented, which is irrelevant if we think that what matters is potentiality itself.

Killing an Innocent Is Always Impermissible

According to this third argument against abortion, abortion is immoral because it is the killing of an innocent human being. Being human confers a right not to be killed, because humans occupy a special place among all other animals (and within this perspective, religious views tend to be, on average, more anthropocentric than secular ones). It is along this line of reasoning that we find views claiming that the killing of an innocent human being is *always* morally wrong.

At first glance, it may seem a good heuristic to say that one should *never* kill any human being, because it is plausible that killing them will cause them pain or deprive them of the opportunity to experience even the simplest form of pleasures. Such a rule, however, carries some odd implications. It means one would have to claim that it is never permissible to kill an innocent human being even if, by not killing him or her, one would cause the death of a million people. Imagine a scenario in which a newly opened EC facility contains only two preserved embryos. One day, workers discover that one of the embryos is infected with an extremely infectious and deadly virus. Worse yet, the cryogenic tank in which the embryo is stored has failed catastrophically, and it is only a matter of minutes before the tank ruptures, releasing the deadly virus on to a nearby city with 1 million inhabitants. The only way to prevent the outbreak and the death of a million people is to burn down the whole cryonics facility, without being able to get back in and rescue the other cryopreserved embryo. It would seem that the right thing to do is to set the cryonics facility on fire, thereby saving 1 million people at the cost of killing two embryos. If one had to claim that killing a human being is always impermissible, they would have to accept the fact that their principle would imply not setting the facility on fire, thereby indirectly killing 1 million people.

Now, obviously, the principle that it is wrong to kill an innocent human being is only a prima facie one, and not an absolute and *all things considered* principle. Although a few people may bite the bullet and accept all of the bizarre implications of the rule that killing a human being is *always* impermissible, many pro-life advocates would ultimately agree that it is permissible to kill a human being under some (perhaps a very limited number of) circumstances. However, abortion is never one such circumstance in the pro-life perspective.

Regardless of which version of the wrongness of killing principle one endorses, the arguments against abortion that we have fleshed out are based on the shared view that killing the embryo/foetus entity would be impermissible because it would cause them to die. But it seems that these same arguments would not justify opposing cryosuspension of pregnancy. Cryosuspension does not cause the death of the embryo/foetus; it only causes them to spend a certain amount of time in a state of nonexistence (or pre-existence). However, as we saw in the third chapter, nonexistence is bad when it is irreversible *and* when it frustrates the desire to continue living, so it would not be bad in the case of an embryo/foetus waiting to be reimplanted. So unless there is some other reason to think that pausing the process that brings someone into existence causes them any harm, then there is no reason to think that cryosuspension of pregnancy would be morally impermissible and that it would not be a better alternative to abortion.

Reproductive Technology

Concerns about the immorality of abortions are not the only reason why some people might feel uncomfortable with the idea of having an abortion. One might think that having an abortion is morally permissible, or not even morally relevant, and yet they might still prefer not to have an abortion. One might feel sadness, regret, distress, frustration about having to make a choice that has irreversible consequences, even when the chosen option is the less bad option between the two available. Cryopreservation of embryos and foetuses would spare women and couples the distress of having an abortion that they might consider moral but that they feel uncomfortable with. This distress can be even more intense if they feel forced by the circumstances to terminate the pregnancy.

At least in some cases, people who have an abortion are not in principle against the idea of becoming parents, but they feel that they are not ready to take up that responsibility at that precise moment of their life (Finer, Frohwirth, & Dauphinee, 2005). They might have only a short-term job contract, and they might be concerned about lack of resources needed to raise a child. Or they might feel unprepared to take on parental responsibilities due to their young age. They might be worried about lacking the psychological stability required to be a good parent. It might be that in a few months, or in a few years, their economic, social, family, or emotional situation will be different and they will want to have a child.

It is because of considerations about the impact of parenthood on people's life goals, and about what constitute good enough circumstances under which people should have children, that reproduction has increasingly become medicalized, and it has changed from a matter of chance to a matter of choice. On the one hand, there are contraceptives that prevent people from getting pregnant at the wrong time or with the wrong person, and, on the other hand, there are infertility treatments that help people to conceive when they feel that there are ideal circumstances to raise a child, but there are biological impediments to conceive.

As people age, it becomes increasingly difficult to conceive a child through natural means, and even more to conceive a healthy child through natural means. Women's fertility starts to decline in their late 20s and early 30s, and by the time they are in their 40s, conceiving a child through natural means is difficult. But most people are focused on their careers, life projects, and finding the right person to start a family during the first three or four decades of their life. Contrary to what used to be the case in the past, the peak of women's fertility no longer necessarily matches the time they feel that it is good for them to become mothers. This is why fertility treatments are so popular, and why many women struggle to conceive.

Given the limits currently imposed by human biology and the current incapacity of medicine to radically extend female fertility, it seems that cryosuspension of embryos (or eggs) would be a good strategy to make sure that women are as free as men to decide whether and when to become parents. Therefore, cryosuspension should be welcome also in a feminist perspective.

Cryopreservation of eggs at a young age would make it possible to utilize these eggs for IVF treatments later on in life, when one has found a partner they want to have children with. Unfortunately, at the moment cryopreservation of eggs has not given as good results as cryopreservation of embryos, so it would be safer to make and cryopreserve embryos with the chosen partner as soon as possible, and then implant them when it is more convenient. If it became feasible to extract and cryopreserve embryos that were accidentally conceived, then one could keep them in an embryo/foetus cryobank and reimplant them at a later time. This way, a woman who got pregnant when she could not continue the pregnancy, could "save up" her own embryo/foetus for a time when she actually desires to have a child, something that could spare her the distress of dealing with infertility treatments later on in life and, perhaps, of having to choose whether or not to do something they find immoral or distressing.

Therapeutic Aid

In some cases, the choice to have an abortion has to do with the health conditions of the foetus. For instance, if tests during the pregnancy reveal a severe genetic condition of the foetus, parents might decide to have an abortion in order to spare their future child the suffering that such genetic condition would inflict on them and/or to spare themselves the difficulties of raising a child with disabilities.

Moreover, even if parents decided to continue the pregnancy knowing that the foetus has a disability, sometimes they do not have the option to continue. Some conditions are not compatible with life; hence the foetus is naturally expelled by the body through a miscarriage.

Both in cases where the parents choose to have an abortion because the foetus has a disability and in cases where the foetus condition causes a miscarriage, cryopreservation of the foetus could be a valid alternative to abortion and miscarriage.

Hopefully, many or at least some of the conditions that are currently disabling or incompatible with life will at some point be curable. Thus, cryosuspension would give prospective parents the option to cryopreserve the embryo/foetus (in cases where the condition causes a miscarriage, they should intervene very early on) and hope that a therapy will be developed while they are still alive, so that they can actually reimplant the foetus and bring it into existence. For foetuses affected by conditions incompatible with life, this kind of intervention would represent the only hope to develop into full human beings at some point in the future. For foetuses affected by disabling conditions that would be compatible with life, cryosuspension would provide a means to buy time in the hope that medicine finds a therapy. If parents decide to reimplant them after a few years because no therapy has been developed and they think that a life with disability is better than no life at all, then cryosuspension would at least have provided such families with more time to learn how to help their future child to have a fulfilling life while carrying a disabling condition. Families often find it difficult to deal with disabilities, to learn how to best take care of a child with special needs, and to organize their home in a way that best accommodates such needs. Keeping the foetus cryopreserved for a while would give them time to prepare for the additional challenges of parenting a child with disabilities.

So it seems that cryosuspension of pregnancies of foetuses with disabilities would be useful regardless of what condition is affecting the foetus and of whether a treatment is developed for such conditions. Of course, it may be that no therapy will ever be found, so that in the end the cryopreserved embryos will never be reimplanted, or, if reimplanted, they will have to live with a disability. But it is plausible that at least for a certain percentage of these embryos that would otherwise be aborted or lost in a miscarriage, cryosuspension would provide a useful aid in providing them with a better life, or with a life at all. Also, in this case, then, cryosuspension is an option that the pro-life should welcome.

Adoption

In cases where people do not want to or cannot have an abortion (they might live in a country where it is illegal), but they also cannot or do not want to raise a child, the only option they have is to complete the pregnancy and to give the child up for adoption. This is often considered an ideal solution because it does not involve killing a foetus and depriving it of a future life, it does not force someone to raise a child under suboptimal conditions, and it allows people who cannot have their own genetically related children to adopt a child and satisfy their desire for parenthood. Yet, this solution has downsides too.

It has been observed that some women are extremely distressed when giving up their child for adoption. Some of them never overcome this traumatic experience, and their psychological well-being is forever compromised (Condon, 1986). Unfortunately, there is also data pointing to the fact that children given up for adoption and passed through different institutions or foster care homes before being adopted (some of them are never adopted) are also negatively affected by this kind of experience (Hoksbergen et al., 2003; Simmel, 2007).

So even though giving up a child for adoption is sometimes considered the best option available in case of unwanted pregnancies, in practice it seems that it can have very negative consequences, and that cryosuspension of pregnancy could be a better alternative. If cryosuspension of pregnancy were an option, people who at the moment can only give up the child for adoption, and as a consequence experience trauma, would have an alternative. They could pause the pregnancy and reimplant the foetus when circumstances are more favourable, for instance, when their economic and personal situation has improved.

But also for women who are confident that giving up a child for adoption would be the best option (they might have a terminal disease, or live in an extremely poor country), cryosuspension of pregnancy could be a useful tool. The pregnant woman could cryosuspend the foetus well before the moment of birth. This way, she could spare herself the burden of nine months of pregnancy, she could avoid to bond too much with the foetus (something that would make the separation less painful), and she would have more than nine months to look for the right adoptive family. This way, she would have better chances to find an adoptive family that matches her preferences. Moreover, an infertile woman adopting the foetus would have the opportunity to have the foetus implanted in her own womb and have time to bond with it at least for a few weeks before it is born.[1] Such a solution would also spare the future child the distress of passing through various foster care homes before finding parent/s that can adopt her.

In sum, cryosuspension of pregnancy could provide a better alternative to adoption both for women who do not want to give their child up for adoption and for the ones who want to. Moreover, it would make the whole process of adopting smoother for both the prospective child and the prospective adoptive family.

Ectogenesis

When considering future technologies that could change the way we procreate, cryosuspension of pregnancy is not the only, or even the first option, that comes to mind. So-called artificial wombs or ectogenesis have been discussed for much longer, and attempts have been made to actually realize such artificial wombs that would allow foetuses to develop outside of a woman's body (Partridge et al., 2017).

In a book published in 1984, Peter Singer and Deane Wells argued that ectogenesis would have helped to solve conflicts related to abortion by allowing unwillingly pregnant women (or women at high risk of miscarriage) to remove a foetus and transferring it to an artificial womb (Singer & Wells, 1984, p. 135). The foetus would have developed in the artificial womb and then, if not wanted by the genetic parents, it could have been adopted by other couples.

Just like cryosuspension of pregnancy, ectogenesis should have constituted a practical solution to bypass moral issues related to abortion, without actually solving such moral conflicts. Singer and Wells thought that pro-life groups would have considered ectogenesis a preferable alternative to abortion, given that it would have saved embryos and foetuses from death caused by abortions (or by miscarriages). At the same time, they expected pro-choice groups to be enthusiastic about a technological means that would have freed women from the burden of pregnancy, giving them unprecedented autonomy and control over their own body.

Singer and Wells argued that ectogenesis should have not deprived women of the opportunity to decide how to dispose of the unwanted embryo or foetus, meaning that abortion should have remained an option. However, if ectogenesis (or cryosuspension of pregnancy) were available, the reasons used to justify abortion could not be ones based on the autonomy of women and their bodily autonomy.

As we saw, there are two main lines of arguments that, either together or separately, could be used to justify abortion. One is based on the view that embryos/foetuses lack significant moral status: there is nothing wrong with killing an embryo/foetus because they do not have capacities or properties that would make them the subject of a right to live. According to a second argument, abortion is permissible because, regardless of the status of the embryo/foetus, a woman has a right to bodily autonomy; hence she cannot be coerced into keeping another human being alive through her own body, regardless of whether such human being has the moral status of a person or not (Thomson, 1976).

If abortion is justified by bodily autonomy, rather than by the lack of personhood of the foetus, then the availability of both ectogenesis and cryosuspension of pregnancy, as ways to keep the foetus alive without breaching a woman's bodily autonomy, would imply that there is no moral reason to justify killing the foetus through an abortion.

However, although it would seem that both ectogenesis and cryosuspension of pregnancy strike a balance between women's right to bodily autonomy (because women will have the possibility not to continue an unwanted pregnancy) and foetuses' right to life (because the foetus will not be killed, but only moved outside the woman's womb), there are some possible downsides that have been taken into consideration. In a book published in 1998, Leslie Cannold interviewed both pro-life and pro-choice women asking them questions about, among other motherhood-related issues, ectogenesis (Cannold, 1998). Perhaps surprisingly, Cannold found that both women justifying abortion on the basis of autonomy and women refusing abortion on the basis of a right to life of foetuses agreed that ectogenesis would not be a morally viable option.

Some of the concerns expressed against ectogenesis would be likely used against cryosuspension of pregnancy. For instance, both pro-choice and pro-life women explained that they would be worried about the possible risks for the development of the foetus if it were transferred to an artificial womb. We can imagine that similar concerns would be raised about the possible consequences of cryosuspension for the development of a foetus. Some of the women interviewed by Cannold seemed to distrust scientists and/or machines, and claimed that they would not feel comfortable relinquishing their foetus to a scientist or to a machine. They were concerned not only about the impact that such procedure might have on the physical development of the foetus, but also about the possible side effects such an unusual "pregnancy" could have on the sense of self and on the overall psychological well-being of the future child well after they were born. In this respect, however, the advantage of cryosuspension of embryos and foetuses over ectogenesis is that the foetus would develop in a uterus. It would form in the woman's body, develop in the uterus for a certain number of days or weeks, would be removed and cryosuspended, and eventually reimplanted in the uterus, where it would resume developing as it happens with natural pregnancies.

Another recurring issue seemed to be that of responsibility. Some women thought that, if a pregnant woman cannot take care of her future child, she should make a decision that proves that she is anyway taking responsibility for this foetus, instead of abandoning it to an uncertain future. Pro-choice women thought that instead of relinquishing it to another person, or even worse, to an artificial womb (or a cryonics facility, in our case), she should make sure the foetus does not develop into an actual baby and does not come into existence. So to pro-choice women, responsibility towards an unwanted foetus is best expressed through the choice of having an abortion, since the alternative would be an uncertain future.

Some women who opposed abortion, instead, thought that ectogenesis was not a good option because it would have given women the wrong belief that, by not killing the foetus and by transferring it to an artificial womb, they were not harming it. However, refusing to take on the responsibility of raising the prospective child would be wrong, just like, they said, it would be wrong to have an abortion. To some women, the wrongness of abortion is not explained exclusively by the harm one (supposedly) inflicts on the foetus by killing it, but also by the fact that a woman refuses to take on herself the responsibility of raising her own child. So, to pro-life women, responsibility towards an unwanted foetus is best expressed through the choice of having that child anyway.

Thus, it would seem that the assessment of the morality of abortion (and of alternatives such as ectogenesis and cryonics) does not rely exclusively on the moral permissibility of killing a foetus, but also on disagreements about how women should express their responsibility towards a foetus that, regardless of their intentions and desires, is growing in their womb. To some people, being responsible for the foetus implies interrupting its development through an abortion rather than bringing them into a world under suboptimal circumstances, including by relinquishing them to a machine or to an adoptive family. To some other people, instead, responsibility towards an unwanted foetus is shown through accepting the pregnancy and raising the child without shifting the responsibility towards someone else.

Also in light of this disagreement about the meaning and the implications of responsibility towards the foetus, it seems that cryosuspension of pregnancy would be preferable to both abortion and ectogenesis. The foetus would not be killed, as in the case of abortion, but would not be relinquished to some machine and then to strangers (unless the foetus were reimplanted in the uterus of the adoptive mother), as in the case of ectogenesis. Thus, the pregnant woman would not refuse to take the responsibility of raising the prospective child—she would only postpone this moment. So irrespective of disagreements about what taking responsibility towards a foetus entails, it seems that cryosuspending the foetus would not imply refusing to take such responsibility on.

Limits

I have considered the cases of women choosing to have an abortion because they get pregnant at a time when they are not able to take care of their prospective child, either because they have projects that are not compatible with parenthood or because they are in a suboptimal psychological or economical state. I have argued that, if given the option to pause the pregnancy for months or years, these people would perhaps choose this option. Hence, they would benefit from having the option of postponing the moment when they become parents without having to kill the foetus they are pregnant with. Similarly, women pregnant with foetuses with disabilities could find cryosuspension of pregnancy advantageous because this pause would give science the time to find a therapy or, at least, it would give prospective parents more time to prepare for welcoming a child with special needs.

However, bad timing and disabilities are not the only reasons why people choose to have an abortion. Some people choose to have abortions because they have no desire to become parents at any time during their life. Some people have abortions because they share David Benatar's view that it is morally impermissible to bring someone into existence (see Chap. 3). Some other people seek an abortion because they do not want to become the parents of that specific child that was conceived with a specific person (e.g. the person who raped them).

In some of these cases, it seems that cryopreservation of the embryos/foetuses could be a good option. Foetuses growing in the womb of women who do not want to have children could be cryopreserved and adopted by people with infertility issues. Couples having problems to conceive could implant these embryos or foetuses (or find a surrogate mother), and fulfil the desire to have a child and to experience a pregnancy. The embryo/foetus would not be destroyed or kept cryopreserved indefinitely, so it would not be deprived of a future life and of the chance of being born in a family that actually wants a child.

However, cryosuspension of pregnancy might not be a good option in some other cases. For instance, if the pregnancy is the result of an abuse, then it is possible that a woman would want to have an abortion in order to eliminate all traces and memories of the violence she was subject to. Even if the woman were reassured that the foetus would not be reimplanted before a very long time, say a century, to make sure that she would be dead by the time the foetus is brought into existence, there is a possibility that the mere existence of such foetus, even though in a cryosuspended state, would make her uncomfortable.

In other cases, a woman could desire to have her embryo/foetus eliminated and not cryopreserved not because the pregnancy is the outcome of an abuse, but because she thinks that bringing someone into existence always equates harming them. In these cases, elimination of the embryo/foetus would be the best option, even better than cryosuspension, other things being equal.

Some people might argue that the interest of some women in having a cryopreserved embryo/foetus donated to them trumps the interest of another woman in not having her embryo/foetus implanted in someone else's womb against her preference. Therefore, in this view, the availability of cryosuspension would always make abortion impermissible, at least as long as there are women who want to have embryos and foetuses of other women reimplanted into their own womb. This would be similar to arguing that, insofar as there are people wanting to adopt a child, abortion is impermissible. However, this view would be quite difficult to defend, as it would be based on an obligation to give something (be it a kidney, a foetus, or money) or to do something to benefit someone (the adoptive family, or the prospective child) even if it goes against one's preferences. We might say that it is morally praiseworthy to donate blood, kidneys, bone marrow, money, and unwanted embryos to people who need them, but respect for autonomy requires allowing people to refuse to do something that to them is extremely burdensome or simply wrong. There are people who do not feel distressed about donating their sperm, eggs, or giving up for adoption a child after they have given birth to them, and these people should certainly be encouraged and praised for their generosity. But it would be unreasonable to expect everyone to have identical emotional responses when giving up their own foetus. So even if cryosuspension of pregnancy became an option, just like donation of kidney is an option, nobody should be forced to donate their embryo or foetus to another person if they feel that such a choice would be detrimental, and therefore no one should be prevented from having an abortion if abortion is what they want.

Cases like the ones just discussed show that cryosuspension of pregnancy would not solve the moral disagreements about abortion, but would only be a practical tool to circumvent some conflicts caused by such disagreements. Even if cryosuspension of pregnancy became a real possibility, there would still be disagreements about what right should prevail: the right to decide on whether or not to have a biological child or the right to life of an enbryo/foetus. But to say that ethical disagreements would not be solved does not mean that cryosuspension would not be a very useful tool for both empowering women and avoiding certain moral conflicts.

It seems therefore that, given the alternatives, there are very good reasons for conducting research on the feasibility of such hypothetical technology. After all, cryonics and indefinite life extension are not available yet, but at some point someone started to think about them before anyone else, and then research started to make cryonics and life extension a real option. Time will tell if cryonics, indefinite life extension, and cryosuspension of pregnancy will ever become available, or if they were just good ideas bound to remain ideas. All the great inventions that make our life much more comfortable than the lives of our ancestors did not appear out of thin air, but are the result of trial and error. We should be grateful to all the people who tried to make the future a more comfortable place, for it is thanks to their successes and failures that progress is achieved.

We cannot predict the outcome of cryonics-related projects, but at least we have reasons to think, or at least I hope I have provided sufficient reasons to think, that they are worth a shot.

Note

1. Some women do not have a womb and would thus not benefit from this; however, many people struggling with infertility would actually be able to carry a pregnancy once they had a healthy embryo/foetus implanted in their uterus. Some people are infertile because of the lack of fertile eggs (usually due to ageing), or poor quality of the sperm. Miscarriages are often explained by congenital defects of the embryo/foetus that is then expelled by the body.

References

Cannold, L. (1998). *The abortion myth: Feminism, morality and the hard choices women make*. St. Leonards: Allen & Unwin.

Condon, J. T. (1986). Psychological disability in women who relinquish a baby for adoption. *The Medical Journal of Australia, 144*(3), 117–119. Retrieved from https://www.ncbi.nlm.nih.gov/pubmed/3945198

Finer, L. B., Frohwirth, L. F., & Dauphinee, L. A. (2005). Reasons US women have abortions: Quantitative and qualitative perspectives. *On Sexual and ...* Retrieved from http://onlinelibrary.wiley.com/doi/10.1111/j.1931-2393.2005.tb00045.x/full

Ford, N. M. (1991). *When did I begin?: Conception of the human individual in history, philosophy and science*. Cambridge University Press. Retrieved from https://market.android.com/details?id=book-VKq7xWqr8g0C

Giubilini, A. (2012). Abortion and the argument from potential: What we owe to the ones who might exist. *The Journal of Medicine and Philosophy, 37*(1), 49–59. https://doi.org/10.1093/jmp/jhr053

Giubilini, A., & Minerva, F. (2012). After-birth abortion: Why should the baby live? *Journal of Medical Ethics, 39*(5). https://doi.org/10.1136/medethics-2011-100411

Hoksbergen, R. A. C., ter Laak, J., van Dijkum, C., Rijk, S., Rijk, K., & Stoutjesdijk, F. (2003). Posttraumatic stress disorder in adopted children from Romania.

The American Journal of Orthopsychiatry, 73(3), 255–265. Retrieved from https://www.ncbi.nlm.nih.gov/pubmed/12921206

Marquis, D. (1989). Why abortion is immoral. *The Journal of Philosophy, 86*(4), 183–202. Retrieved from https://www.ncbi.nlm.nih.gov/pubmed/11782094

Minerva, F., & Sandberg, A. (2015). Cryopreservation of embryos and fetuses as a future option for family planning purposes. *Journal of Evolution and Technology/ WTA, 25,* 17–30. Retrieved from http://jetpress.org/v25.1/minerva.htm

Partridge, E. A., Davey, M. G., Hornick, M. A., McGovern, P. E., Mejaddam, A. Y., Vrecenak, J. D., ... Flake, A. W. (2017). An extra-uterine system to physiologically support the extreme premature lamb. *Nature Communications, 8,* 15112. https://doi.org/10.1038/ncomms15112

Pavone, M. E., Innes, J., Hirshfeld-Cytron, J., Kazer, R., & Zhang, J. (2011). Comparing thaw survival, implantation and live birth rates from cryopreserved zygotes, embryos and blastocysts. *Journal of Human Reproductive Sciences, 4*(1), 23–28. https://doi.org/10.4103/0974-1208.82356

Simmel, C. (2007). Risk and protective factors contributing to the longitudinal psychosocial well-being of adopted foster children. *Journal of Emotional and Behavioral Disorders, 15*(4), 237–249. https://doi.org/10.1177/10634266070150040501

Singer, P., & Wells, D. (1984). *The reproduction revolution: New ways of making babies.* Oxford: Oxford University Press.

The World's Abortion Laws Map. (2014). Retrieved February 15, 2018, from https://www.reproductiverights.org/document/the-worlds-abortion-laws-map

Thomson, J. J. (1976). A defense of abortion. In J. M. Humber & R. F. Almeder (Eds.), *Biomedical ethics and the law* (pp. 39–54). Boston, MA: Springer. https://doi.org/10.1007/978-1-4684-2223-8_5

Index[1]

A
Abortion
  embryo, 6, 17, 96, 112–114, 120, 122, 125, 128, 129
  foetus, 6, 96, 112–114, 120, 122–129
  voluntary cryosuspension of pregnancy, 96, 112–120, 122–129
Amortality, *see* Immortality

B
Benatar, David, 51, 52, 113, 128
  Better never to have been, 51–52
Brain
  neural substrate, 11
  neuropreservation, 25, 26
  uploading, 11, 42n1, 68, 84
  tumour, 118

C
Consciousness
  loss of, 9
  mind, 9, 10
  neural information, 8–10
Cryonics providers
  Alcor, 18, 25
  Cryonics Institute, 18, 20n2, 28, 29
  Kriorus, 25
Cryopreservation
  of adults, 17
  of brains, 25
  of embryos, 117, 120, 121, 128
  of euthanized patients, 96, 111
  of foetuses, 6, 96, 113, 114, 116, 117, 120, 122, 128
Cryothanasia, *see* Euthanasia

---

[1] Note: Page numbers followed by 'n' refer to notes.

© The Author(s) 2018
F. Minerva, *The Ethics of Cryonics*,
https://doi.org/10.1007/978-3-319-78599-8

D

Death
  cardiopulmonary standard, 9
  information-theoretic criterion, 9–12, 104
  whole-brain standard, 9
Doctrine of double effect (DDE), 101, 102

E

Embryo cryopreservation (EC), 5–7, 12, 16, 17, 113, 119
Ettinger, Robert, 12
  The prospect of immortality, 12
Euthanasia
  assisted suicide, 98
  cryothanasia, 7, 96–105, 107, 108, 111

F

Fallacy
  naturalistic, 13, 14
  status quo bias, 53
  sunk cost, 53
Fischer, John Martin, 65, 75, 78, 86, 113
  Immortality, 75, 78
  Immortality and boredom, 65, 86

G

Gard, Charlie, 30
Genetically modified organism (GMO), 16
Giubilini, Alberto, 98, 113, 115, 118
  Abortion and the argument from potential, 115
  Euthanasia: what is the genuine problem?, 98

H

Heidegger, Martin, 78, 113
  Being and time, 78

I

Immortality
  chosen, 68
  coerced, 68, 83
  indefinite life extension, 45, 61, 63, 65, 68–70, 82, 83, 86, 92
Intensive care unit (ICU), 15, 30, 32, 85
In-vitro fertilization (IVF), 4–7, 12–19, 33, 59, 81, 106, 108, 113, 121

K

Kagan, Shelly, 86, 113
  Death, 86
Kass, Leon, 113
  Why we should ban human cloning now, 15

L

Liquid nitrogen, 4, 7–9, 13, 15, 17, 24, 38, 108
  cryopreservation, 4, 8

M

McMahan, Jeff, 9, 57, 58, 64, 113
  Death and the value of life, 58, 64
  Metaphysics of brain death, The, 9
May, Todd, 71, 79, 87, 113
  Death, 71, 79, 87
Mill, John Stuart, 13
  nature, 13

Mitchell-Yellin, Benjamin, *see* Fischer, John Martin
Moen, Ole Martin, ix, 99, 113
The case for cryonics, 99–100

N
Nagel, Thomas, 50, 53–57, 60, 61, 115
Death, 50, 54, 57

O
Organ donation, 25, 26

P
Paralysis
 locked-in syndrome, 9
Parfit, Derek, 76, 77, 113
Reasons and persons, 76

R
Rejuvenation, 36, 45, 46, 61, 63, 69, 78, 82, 84, 86
Religion
 Catholic Church, 17, 20n5, 101
 God, 1, 7, 14–15, 17, 19, 68, 84, 85, 103
 Heaven, 17, 68, 83–86
 Hell, 68, 84, 86

Playing God, 15
Soul, 15, 17, 20n5, 67, 68, 83, 84, 93n4

S
Sandberg, Anders, x, 8, 99, 107, 113, 114
Cryopreservation of embryos and fetuses as a future option for family planning purposes, 8, 113
Euthanasia and cryothanasia, 97, 107
Scheffler, Samuel
Death and the afterlife, 79
Singer, Peter, 12, 90, 114, 124, 125
In vitro fertilisation: the major issues, 12
Life's uncertain voyage, 90

V
Virtual reality, 26

W
Williams, Bernard, 62–65, 74–78, 86, 114, 118
Makropulos case, The, 62
Problems of the self, 62

CPI Antony Rowe
Eastbourne, UK
February 15, 2019